MW01517676

EXCAVATING
VICTORIANS

Suny series, Studies in the Long Nineteenth Century
Pamela K. Gilbert, editor

EXCAVATING VICTORIANS

Virginia Zimmerman

STATE UNIVERSITY OF NEW YORK PRESS

Cover: Untitled geological still life, frontispiece to Gideon Mantell's *The Medals of Creation* (1844). Courtesy of Special Collections, University of Virginia Library.

Published by
State University of New York Press, Albany

For information, contact State University of New York Press, Albany, NY
www.sunypress.edu

Production by Marilyn P. Semerad
Marketing by Michael Campochiaro

Library of Congress Cataloguing-in-Publication Data

Zimmerman, Virginia.
   Excavating Victorians / Virginia Zimmerman.
     p. cm. — (Suny series, studies in the long nineteenth century)
   Includes bibliographical references and index.
   ISBN 978-0-7914-7279-8 (hardcover : alk. paper)   1. English literature—19th century—
History and criticism.   2. Geology in literature.   3. Archaeology in literature.   4. Space
and time in literature.   5. Time—Philosophy.   6. Authors, English—19th century—
Philosophy.   7. Literature and science—Great Britain—History—19th century.
8. Literature and history—Great Britain—History—19th century.   I. Title.

PR468.G38Z56   2008

820.9'008—dc22   2006101369

                                        10 9 8 7 6 5 4 3 2 1

Contents

Illustrations

Acknowledgments

For permission to reproduce selected images, I would like to acknowledge with thanks: the Guildhall Library, City of London (figures 4.1 and 4.5), and the Ministero per i Beni e le Attività Culturali-Soprintendenza Archeologica di Pompei (figures 4.3 and 4.4). Special Collectons at the University of Virginia library made available the frontispiece to Gideon Mantell's *The Medals of Creation*, which appears on the cover of this book, for which I thank them. In addition, I am grateful to David Latané and *VIJ: Victorians Institute Journal* for permission to use in chapter 2 a revision of my article published there in winter 2002.

I would like to acknowledge the support, guidance, and kindness of James Peltz, Allison Lee, and Marilyn P. Semerad at the State University of New York Press, and of course I am also grateful to series editor Pamela K. Gilbert.

I am grateful to Beth Cunningham and Eugenia Gerdes at Bucknell University for generous grants of funds to travel to England, where I visited libraries, parks, and ruins, all of which have found their way into this project, and for the even more precious gift of time in the form of untenured faculty leave. Also at Bucknell, I would like to acknowledge the unflagging support and kindness of the members of the English department who, formally and informally, readily engaged with this work as it progressed. I am especially grateful to Michael Drexler, Meenakshi Ponnuswami, Greg Clingham, Harold Schweizer, and John Rickard for listening kindly through the latter stages of this project. I would also like to thank Chris Camuto and Eric Faden whose work reminds me that I am not isolated in what I do.

I owe a particular debt to Saundra Morris for keeping me honest about how I spent my time, and most especially to Ghislaine McDayter for answering innumerable questions, being an ever-patient sounding board, and most happily sharing in some much needed time away from work. In addition, I would like to thank graduate students Sean Martin, Ed Kelleher, and Rebecca Morris whose assistance made my life easier and my book better.

I remain grateful to Alison Booth, Karen Chase, Michael Levenson, and Herbert Tucker at the University of Virginia. Each of them saw this project in its earliest stage, and through many readings, generous feedback, and lively conversation did much to elevate my work. I especially appreciate their continued interest in this project. Other members of the University of Virginia community were kind and generous over the years, notably Stephen Cushman, Steve Arata, and Kay Neeley.

There would be no *Excavating Victorians* without the years of support and readings of early drafts offered by Amanda French, June Griffin, Elizabeth Outka, and Lisa Spiro. For continued friendship and intellectual support, I am indebted to John Picker, who has helped me shape my ideas and has counseled me in how best to share them. I would especially like to acknowledge Michelle Allen who has been my friend and colleague from the beginning and has offered everything from perfect phrasing to stimulating conversation and from diligent reading even at the busiest of times to top-notch god-parenting. Additionally, I am deeply grateful to Constance Walker and the faculty of the English department at Carleton College; they challenged me to read more deeply than I knew one could and prepared me to ask the questions that led me to excavate the Victorians.

For reading, conversation, and other help various and sundry, I would like to thank Robert D. Aguirre, Lisa Perrone, Andrew M. Stauffer, Joe McLaughlin, James Secord, and Gillian Beer.

For actually taking me fossil hunting and for checking my geology, I am indebted to good friends Debbie Ware and Jo Fleming, who have the good luck to live in Cuckfield.

I would especially like to acknowledge the unwavering interest and support of my family and friends: Michael Zimmerman, Mary Zimmerman, Kristen Fennel, Marion Von Beck, Frank Comas, and all my family in Barcelona, who witnessed the inception of this book.

I am most grateful of all to Jordi Comas, whose unflagging faith in my work has made the work so often a pleasure. He has traveled with me and read with me throughout this project, and he has asked hard questions but never questioned me. The traces of our conversations and explorations have shaped the landscape of this book, and the journey has been most enjoyable because of his companionship.

Finally, I must acknowledge Elijah, Alexander, and Dorothea, my personal paradigm shifts, who have inspired me to appreciate time and traces anew.

1

Introduction

"All Relics Here Together"

In a paper titled "Theory of the Earth" (1788), geologist James Hutton (1726–1797) wrote that in his study of the Earth's development he found, "no vestige of a beginning,—no prospect of an end" (304).[1] The nineteenth century inherited this terrible legacy, a new sense of a boundless time that ultimately replaced the Bible's compelling narrative of time's very clear beginning and no less certain end. This familiar plot, aside from its theological import, framed time and stressed the centrality of human life. Geology, in contrast, burdened the nineteenth century with a sense of time that exceeded the limits of plot. Throughout the nineteenth century, writers struggled to conceive the infinity evinced by the natural world, a world newly understood as indifferent to human experience. Writing in the *Quarterly Review* in 1827, geologist Charles Lyell (1797–1875) states the case plainly: "all discoveries which extend indefinitely the bounds of time must cause the generations of man to shrink into insignificance and to appear, even when all combined, as ephemeral in duration as the insects which live but from the rising to the setting of the sun" (474). A generation later, Alfred Tennyson mourns the loss of a kinder nature that valued human life in *In Memoriam* (1850): utterly bereft, his speaker likens himself to "an infant crying in the night," and indifferent nature coldly responds, "I care for nothing, all shall go" (54, 56). Tennyson articulates what many had realized by mid-century: "all" describes whole species and inevitably includes humanity. Fossilized creatures that lived and died eons past were messengers from the depths of geological time, revealing at once the immensity of the past and the certain fate of humans in the future. Writing of Hutton in volume 1 of *Principles of Geology*, Lyell explains: "Such views of the immensity of past time . . . were too vast to awaken ideas of sublimity unmixed with a painful sense of our incapacity to conceive a plan of such infinite extent" (63). Yet, Lyell's project, one shared by many Victorian writers, was to conceive the infinite plan.

Through geology—and archaeology, which developed in the young science's wake—scientists fashioned narratives out of fragmented remains. The evidence they excavated revealed at once the extraordinary depth of time and the awesome ability of the writer to measure time and to craft its story. Immediately after marveling at the immensity of time and the difficulty of conceiving such an elaborate and infinite plan, Lyell asserts the capacity to imagine boundless time: "Worlds are seen beyond worlds immeasurably distant from each other, and beyond them all innumerable other systems are faintly traced on the confines of the visible universe" (63). Though time and space may be immeasurable, distant worlds are "seen" nonetheless. A consideration of Lyell's strategic use of passive voice reveals the paradox that empowered geology and archaeology in the face of the "immensity of past time": in this sentence, it is ambiguous who or what has performed the action ("traced"), and thus Lyell invokes at once the natural processes that leave evidence of change on the "visible universe" and also the scientist who interprets the traces of the past writ on the landscape. This conflation is echoed throughout the writings of the nineteenth century: the infinite expanse of time is matched with the geologist whose careful calculations make time vast, indeed, yet finite; the oppressive layers of dust that entomb not just individuals but entire cultures are plumbed by archaeologists whose discoveries give life to the very cultures and individuals that seem to have been destroyed.

Again and again, geologists, archaeologists, and writers influenced by these new sciences offer in one image, in one sentence, both the destruction and the preservation of the individual. The writer is empowered to understand and explain the "immensity of past time," and this position as narrator of time's vastness rescues the individual, in this instance the writer, from being obliterated by time. In *The Sense of an Ending*, Frank Kermode describes world history as "the imposition of a plot on time" (43). Natural history is somewhat different. As Gillian Beer asserts in *Darwin's Plots*, "Lyell, and later Darwin, demonstrated in their major narrative of geological and natural history that it was possible to have plot without man—both plot previous to man and plot even now regardless of him" (17). The Earth has its own plot, not imposed by a person or a divinity. However, it requires the imagination of human observers to discern that plot and read it out. Beer later explains,

> The humanistic core of Lyell's work is its insistence on the power of man's imagination, which allows him to recuperate the staggeringly extended time-scale of the physical world. Though his presence is diminished in the raw time-scale, his is the only source of powerful *interpretation*. (Beer's emphasis; 39)

This notion of "powerful interpretation," or even empowering interpretation, is the subject of this book. How does the individual look into the abyss of time and escape annihilation? Beer argues, and I emphatically agree, that narrative offered a way to imagine at once time's expanse and the persistent value of the individual life.

Indeed, the authority of the excavator to tell the story of the Earth, a narrative constructed from the vestiges of the past, privileges the present and thus diminishes the alarming implications of the abyss that is the past.[2] It is from his or her position grounded in the present that the geologist or archaeologist examines fossils or ruins and draws conclusions about the past; he or she might also make predictions about the future. In both cases, the other times are dwarfed by the here and now, the present looming larger than the expansive past in its actuality, if nothing else. In her analysis of the human experience of time, *Between Past and Future*, Hannah Arendt describes the privileged position of the present: she writes that past and future have no known beginnings and both terminate in the present, a variation on Hutton's theme. From this vantage point at the middle of time, writers assert considerable authority over the past and the future; as they muse on what has been and what will be, they always emphasize their own moment. Walter Benjamin similarly stresses the paramount importance of the present in history in his "Theses on the Philosophy of History"; he writes, "History is the subject of a structure whose site is not homogenous, empty time, but time filled by the presence of the now" (263). Benjamin uses a spatial metaphor, as so many who write about time inevitably do, and describes the present as being able to fill up all the space of history: thus, the present, despite all its fleeting qualities, is physically equivalent to all of the past. In the insistence on the importance of the present as a measure of both past and future, individuals claim a position of spatial, temporal, and narrative significance.[3]

In the chapters that follow I discuss the literature of excavation, an activity founded on the fundamental relationship between time and space: an object is removed from layers of Earth that signify the passage of time and re-presents the past. I pay particular attention to narratives from geology and archaeology, as well as literary texts that echo the concerns of these sciences. Born in the late eighteenth century, geology grew into a premier science in the first third of the nineteenth century. Like many sciences at this time, geology was accessible to everyone; there were no specialists, and many men and women from all strata of society read the texts published on geology and even participated in excavations. For much of the century, archaeology was regarded as a close cousin or even a branch of geology.[4] Excavation was more geographically than disciplinarily bounded, and a single site often yielded interesting fossils as well as notable remains from human cultures. Some who excavated such sites made no distinction between these finds and happily displayed fossil fishes alongside prehistoric tools and human bones. In fact, the proximity of human remains to extinct faunal remains made the implications of geology for humanity very clear: people and their cultures are no more resistant to the passage of time than are bivalves or dinosaurs.

The relationship between the individual life and the immensity of time was a central concern for early geology, and the struggle between catastrophists and uniformitarians was, in addition to being an important and ongoing scientific controversy, a narrative struggle to define that relationship. Some early scientists interpreted the fossil record as evidence of a great catastrophe, some cataclysmic event that destroyed entire species in one blow.[5] So-called catastrophists more easily assimilated scientific evidence with the traditional biblical account of Earth's history: for instance, they famously conjoined science and theology with their attention to the flood as the central geological event, a compelling explanation for the discovery of marine fossils on mountaintops. Yet, Charles Lyell and his followers promoted a different interpretation: what is now the craggy top of a mountain may once have been a level plain resting beneath the sea. Uniformitarianism, championed by Lyell, assumes that processes in action in the present are identical to those in action in the remote past. So the past can be witnessed by proxy, as it were, by the geologists who examine the processes they can observe in the present—as Arendt remarked, from the vantage point of the present, we can observe all time. In addition to its assertion of the value of the present, uniformitarianism assumes a vastly expanded time scale. Over such a tremendous amount of time, very small actions affect great change. For example, if small pebbles are knocked against the shore again and again for millennia, they will carve a coastline. Uniformitarian geologists believed the movement of the pebbles to be uniform, and they allowed millions of years for the pebbles to do their work; the equation results in awesome change caused by individuals who had formerly appeared to be absolutely insignificant. Thus, even as it exposed the terrible specter of eventual annihilation, geology offered a theory that empowered the small, the individual.

Matthew Arnold (1822–1888) explores the impact of geology on the individual in his poem "Dover Beach" (1867).[6] He breaks the calm he establishes in the first few lines of the poem—"The sea is calm to-night" (1)—with the imperative in the ninth line, and he continues on to describe the powerful action of pebbles:

> Listen! you hear the grating roar
> Of pebbles which the waves draw back, and fling,
> At their return, up the high strand,
> Begin, and cease, and then again begin,
> With tremulous cadence slow, and bring
> The eternal note of sadness in.
>
> (9–14)

Arnold here describes the "grating roar" of uniformity: the small actions of the pebbles he hears in the present echo the same pattern occurring day after day for countless years. The "tremulous cadence slow" describes not only the sound of the rocks but also the sound of the ages passing slowly, slowly, and Arnold states in line fourteen that they bring "the eternal note of sadness in." His phrasing here is subtle and important: the grating roar does not bring in sadness; rather it brings in the eternal note. The pebbles grating against the strand make a sound that endures across all time; it is eternal. The note is also one of sadness, a sound that depresses rather than elevates the spirit. Indeed, Arnold goes on to describe a different roar: the "melancholy, long, withdrawing roar" of the "Sea of Faith" (25, 21). The note the pebbles sound forces the sea of faith to withdraw, and this withdrawal is also a "roar." By repeating the word "roar," Arnold makes a clear connection: the sound of the pebbles drowns out the sea of faith. Yet, for all his lovely articulation of science-inspired angst, Arnold does offer hope.

The grating roar does not bring in sadness; it only brings in a note, a sound, which may oppress with its fierce and persistent roar, but which has no inherent value. The person who hears the roar may feel sadness, especially if the roar dampens the soothing sounds of faith, but he could also feel empowered. Let us return to the imperative of line nine: "Listen!" The grating roar of the pebbles only signifies the passage of a tremendous amount of time if someone listens to their sound and interprets its meaning. Arnold first compels the reader to hear the note and to drown in its annihilating significance, but he reverses this hopeless sentiment in the final stanza when he offers a second imperative: "Ah, love, let us be true / To one another!" (29–30). When the reader listens at line nine, he or she hears geological time and can only conclude that human lives are worthless, but when Arnold makes this second command, he insists on the value of an individual life. The interlocutor is exhorted to value another person despite the "darkling plain" of life, rife with confusion and ignorance. Love and the bond between two people, despite all evidence to the contrary, matter in this poem. Arnold even matches the "eternal note of sadness" with his own note; the "Ah" he sounds at the beginning of the poem's last stanza vies with the depressing note and strives to uplift. The embattled lovers may not be audible above the grating roar of science, but they huddle together and drown out faith's retreat. And Arnold's "Ah" is audible: the poet-speaker makes a sound that resounds alongside the "eternal note of sadness," and the result is a fierce harmony that drowns out the individual voice even as it throws that desperate cry into relief. In the following chapters, I examine texts from geology and archaeology alongside works by Alfred, Lord Tennyson and Charles Dickens that struggle to hear and amplify the individual's cry.

Making Meaning of Fossils

Fossils invited the Victorians to think deeply about time. Similarly, many historians of science have used fossils to structure their histories: Martin J. S. Rudwick charts the discovery of the geological time scale through these remains in *The Meaning of Fossils*, Claudine Cohen traverses the same territory through a single fossil-figure—the varied forms of the mammoth—in *The Fate of the Mammoth*, and Simon J. Knell details the development of scientific societies through focus on the fossil in *The Culture of English Geology, 1815–1851*. Each of these histories takes as its subject fossils and how scientists advance their interpretations of them. Others, such as Paolo Rossi and Claude Albritton, have offered more general overviews of the history of the "discovery of time," as Stephen Toulmin and June Goodfield call it. They do not emphasize fossils or any other particular element of geology; instead, they present a time line of individuals whose scientific contributions evolve toward what we consider fact today.[7] Yet, even these more expansive histories of geology remain tied to scientific discoveries, whether analyses of fossils, advances in stratigraphy, or refined understanding of comparative anatomy. Such a focus is entirely appropriate, for these works are histories of science. *Excavating Victorians*, however, is neither a history nor a piece of scientific writing, and thus in its genre and its content it diverges essentially from this body of literature to which it is much indebted.

Science is the subject, sometimes directly and sometimes obliquely, of the writers whose writing is my subject. My focus is on the literature inspired by advances in geology, not on geology itself. In the twenty-first century, considering science and literature together seems novel, an intriguing conjunction invited by academia's recent emphasis on interdisciplinary study. However, the disjunction that makes the conjunction so appealing is itself a product of the twentieth century and obscures the fact that disciplinary boundaries are a recent invention. Science and literature were not distinct from one another in the nineteenth century. Beer states the case very clearly at the outset of *Darwin's Plots*, her great contribution to the literary study of what we now consider to be scientific texts: Darwin and his contemporaries "shared a literary, non-mathematical discourse which was readily available to readers without a scientific training. Their texts could be read very much as literary texts" (4). She goes on: "Moreover, scientists themselves in their texts drew openly upon literary, historical and philosophical material as part of their arguments" (5).[8] She describes how metaphors and narrative patterns moved freely between scientists and nonscientists: for insistence, she argues that Darwin was influenced no less by Malthus's prose than by his theory, and further that he was influenced no less by Milton and Dickens than by Malthus. And Beer makes certain her reader understands that though Darwin is her focus, the patterns she describes are typical of the period.

I follow Beer in attending to the literary discourse that arises out of the science of geology. This body of literature contains works by scientists—I pay particular attention to Lyell and Gideon Mantell (1790–1852)—and more traditional literary figures, such as Tennyson and Dickens. I also examine writings published in the popular press, travel literature, and the reports of professional archaeologists, all commenting on sites of archaeological ruin in London and at Pompeii. To be clear: I posit no causal relationship—Dickens, for one, was not influenced by science writing any more than scientists writing were influenced by Dickens. My subject is a discourse that includes the categories we today identify as "literature" and "science," but that in the nineteenth century was simply a body of writing addressing a changing understanding of the Earth and the value of human life. Beer writes explicitly of her methodology and offers an excellent gloss on my own: "My method pays close attention to Darwin's language. He did not *invent* laws. He *described* them" (Beer's emphasis; 46). She analyzes Darwin's language, his descriptions: in short, her focus is not on his science but on his writing, though she is persuasive in her insistence that the two should not be severed. Like Beer, I attend to text, and I am much influenced by her description of the symbiotic relationship of science and literature. She explains:

> Though the events of the natural world are language-free, language controls our apprehension of knowledge, and is itself determined by current historical conditions and by the order implicit in syntax, grammar, and other rhetorical properties such as metaphor, as well as by the selective intensity of individual experience. (46)

Beer makes the important point that language is important not only because it describes natural phenomena, but also because it serves as the intellectual system that enables an observer to make sense of nature first in his own mind and only later in narrative form. The frequent use of reading as a metaphor for geological study illustrates the point: Lyell, for example, wrote that study of the present-day Earth reveals the "alphabet and grammar of geology," and it is the task of the geologist to close read that terrestrial text. It is my task to close read the texts that arose from the work of scientists such as Lyell and the many others who participated in the discourse of reinventing time.

The Observer and the Trace

When an observer excavates a site and uncovers a fossil or an artifact, she raises that object from the depths of time past to the surface of the present. The excavator, thus, has the power to move objects from one time into another. I mean this literally—a geologist lifts a fossil shell from the strata that preserves other marine life from the oolitic period and removes the shell to a Victorian display case—and also figuratively—an archaeologist extracts a small marble ball from its lava casing at Pompeii, interprets his find, and claims that Roman children played with marbles as Victorian children do. In an essay titled "The System of Collecting," Jean Baudrillard theorizes the life of objects that are removed from their original contexts and reinvented as items in a collection. He asserts, "Once the object stops being defined by its function, its meaning is entirely up to the subject" (7). Geological and archaeological traces are sometimes defined by their function, but they are always also interesting because of what they signify about the past and about the passage of time. Thus, these objects are important because of their revised function as temporal signifiers. As the people who physically control the artifacts and, more importantly, who control their interpretation, excavators wield tremendous power in the face of the expanding time scale—they read the signifiers. As the authority of the Bible waned, the authority of the scientists and authors influenced by geology and archaeology grew. Their authorial power allowed them to tell the story of time, and their story depended on the relationship between the object and its discoverer. Indeed, excavation became a powerful epistemological trope for the Victorians: it brings together notions of time as spatial and of a person's ability to stand outside the layers of the past (that is, the physical evidence of time's passage) to plumb the depths and produce narrative accounts of what is uncovered within the rock and dust. Excavation was a way of knowing and conquering the depth of time.

The "trace," as defined in *Time and Narrative* (1985), by Paul Ricoeur, twentieth-century philosopher of science, is the central object of excavation and the focal point of so many narratives of geology and archaeology. Conveying a mutli-temporality, "trace" indicates an object's origin in the past but also its duration into the present. Ricoeur is particularly interested in the relationship between material things and time. Such things offer a measure not only of time's passage but also its peculiar stasis. Emphasizing its duration, Ricoeur connects the trace to what Edmund Husserl calls a *zeitobjekt*, or "tempo-object" (26). Much like Arnold's grating roar, the *zeitobjekt* is a sound that continues. Ricoeur explains that the "tempo-object" indicates "duration, in the sense of the continuation of the same throughout the succession of other phases" (26)—the pebbles cause a roar that begins, ceases, and then begins again. Thus, traces, for all their connection to the past, are emphatically of the present. This concept distinguishes the trace from, say, an artifact: unlike some other material objects, the trace develops its significance from its duration, that is, from its origin in the past yet its presence in the present. In their study of archaeology, Michael Shanks and Christopher Tilley echo Ricoeur:

> The past is conceived as completed. It is in grammatical terms "perfect," a present state resulting from an action or event in the past which is over and done. This "perfected" past is opposed to the flow of the ongoing, incompleted, "imperfect" present. Although the past is completed and gone, it is nevertheless physically present with us in its material traces. (9)

Ricoeur might rejoin that the fact of traces means that the past, too, is imperfect: despite the many histories that neatly divide time, no period or epoch is really discrete. Simultaneously making the past legible and eroding temporal boundaries, the trace invites both empirical study and imaginative interpretation. It is this last that is the subject of this study.

Of course, the trace most famously entered the discourse on meaning and interpretation in Jacques Derrida's *Of Grammatology*. Derrida explains that the trace signifies an elusive originary something, perhaps a thought. It signifies the loss of origin yet is at once a sign of origin. Thus, the geological trace is both the sign of the deep past (a past that theoretically stretches back toward beginnings and origin) and a sign that the past is passed. Derrida emphasizes the erasure inherent in signification: that is, the trace may suggest the past, but it is inevitably apart from that past and in its very existence reveals the irretrievability of what is signified. I differ from Derrida in my interpretation of Ricoeur's representation of the trace. I believe that Ricoeur's emphasis on endurance amounts to an emphasis on preservation: while Derrida sees meaning always eroding, Ricoeur sees the possibility that meaning can be sustained, though perhaps in a diminished form. The trace is not the past—this I do not dispute—but it is a material connection to the past, and while it shows erasure, it also preserves. Moreover, it becomes the foundation for the new historiography that aims to present the past through narrative.

Historians like Thomas Carlyle and Thomas Babington Macaulay, along with writers of historical fiction such as Edward Bulwer-Lytton, sought to bring the past to life through the use of essentially literary narratives. In his discussion of nineteenth-century historicism, Michael I. Carignan explains:

> The view that history is a branch of literature dominated the Romantic historiography of the 1820s and 1830s, and most of the well-known historians of the period strove to write grand narratives that were appealing as literature. . . . In Macaualay's mind, the historian's talent "bears a considerable affinity to the talent of a great dramatist." (399)

Writing of Jules Michelet, Roland Barthes describes the "historian-as-bone-collector and restorer of human dust" (84). In this sense, the historian works from traces, whether bones or written accounts, and from these traces crafts a narrative that not only makes vivid the events of the past, it also eliminates the pastness from the story, bringing the past back to life in the present. Barthes writes:

> At the heart of every resurrectional myth . . . , there is a ritual of assimilation. The resurrection of the past is not a metaphor; it is actually a kind of sacred manducation, a domestication of Death. The life Michelet restores to the dead is assigned a funereal coefficient so heavy that resurrection becomes the original essence, absolutely fresh and virgin of death, as in those dreams where one sees a dead person living, while knowing perfectly well that the person is dead. (84)

This conjunction of history and resurrection runs through much of the literature of excavation: geology brings Victorians into direct (if fictional) contact with dinosaurs, archaeology reanimates Pompeii, and Dickens literally resurrects Rogue

Riderhood and John Harmon in *Our Mutual Friend*. The past is made to live again, figuratively through narrative, and literally through the removal of its trace into the material spaces of the present.

In the third volume of *Time and Narrative*, Ricoeur emphasizes the substantive quality of the trace:

> Hence the trace indicates "here" (in space) and "now" (in the present), the past passage of living beings. It orients the hunt, the quest, the search, the inquiry. But this is what history is. To say that it is a knowledge by traces is to appeal, in the final analysis, to the significance of a passed past that nevertheless remains preserved in its vestiges. (120)

Central to Ricoeur's definition of the trace is the pastness of the object that the trace signifies. A footprint offers a classic example:[9] a person crossed the beach at some time prior to the present; the person is gone, but her footprint remains. The trace records the fact that someone was there and now is not. Thus, the trace at first suggests a linear model of time: then precedes now. Yet, the trace is an object in its own right, not simply an indicator of the current absence of another object, and it exists in the present. In its dual function as a sign of the past and a thing of the present, the trace seems to exist outside time or across all time. Thus, if history is "a knowledge by traces," as Ricoeur asserts, then history is made from the interpretive exchange between past and present. Carignan connects this temporal exchange with Hegel, whose "philosophy helped organize the idea that the past was organically connected to the present and that its coherence depended on the consciousness of the historian" (399). Barthes makes explicit the historian's power over time in his description of Michelet's view: "the historian is the man who has reversed Time" (83).[10]

Yet, the multi-temporality of the trace is even more complex than Ricoeur posits or than Barthes acknowledges, for it is not merely a lingering object from an earlier era—it is also a function of time's passage. To show what I mean I will extend the example of the footprint: the footprint was left at some point in the past by the foot of a particular person who walked over the sand. It signifies that in the past a person with a foot of particular shape and size moved across the site. With the passage of time, the action of water and wind has caused the footprint to erode; its edges become less distinct and its depth diminishes. Thus, the footprint that we observe in the present is a function of the original action and also the passage of time. As a trace, it offers information about the past but also about the time passed between past and present, and even about the present itself: we can perhaps determine how much time has passed, what the weather conditions may have been over that period, and we integrate the presence of the trace into our view of the present landscape. The trace endures and thus assumes a temporal significance that transcends its original function. As it "re-presents" what went before (Ricoeur 34–36), the trace diminishes the pastness of the past and makes the depth of time visible and comprehensible. Thus, the excavator or historian who writes a history built on a foundation of traces writes about past and present, but also about time itself.

Thomas Huxley (1825–1895) writes in "On the Method of Zadig" (1880) about what he calls "retrospective prophecy," and he offers a compelling reflection on the authority of the individual who crafts a narrative about the past based on material evidence available in the present. "Retrospective prophecy" seems a particularly apt term to describe the work of excavators in the nineteenth century as it emphasizes not only the past but more precisely looking at the past and crafting a narrative. The materiality of the artifact is a key piece of the puzzle, for the excavator can literally look at a trace of a time past. In reference to Zadig, a naturalist from ancient Babylon described by Voltaire in a philosophical treatise written in 1747, Huxley applies the language of seeing into the future to the science of reading the past: he acknowledges that some thought Zadig had extraordinary powers and thus regarded him as a prophet, yet his powers were simply those of rigorously applied scientific method. Huxley explains that Zadig excelled at meticulous observation of detail and, most importantly, he recognized in the small actions of the present the processes of the past: Huxley explains that examination of a pattern of broken twigs enables Zadig to deduce what sort of party passed by (this is a variation on my example of the footprint). Huxley goes on: "For the rigorous application of Zadig's logic to the results of accurate and long-continued observation has founded all those sciences which have been termed historical or palaetiological, because they are retrospectively prophetic and strive towards the reconstruction in human imagination of events which have vanished and ceased to be" (9). A skilled observer like Zadig can build from fragments—traces—a narrative of the past, but that narrative is a product of the present and resides in the imagination of living humans.

The perception that excavation provided a direct encounter with the past led easily to the assumption that the excavator conquered the past, and I will turn now to the nexus of imperialism, new historiography, and the materiality of the trace to explore further this important issue of control, or authority. As the British traversed the globe on imperial missions, they colonized foreign lands and took the alien artifacts they found there as trophies to proclaim their moral, cultural, and military victories over "lesser" peoples. Many of these trophies were objects of geological or archaeological interest, so the acquisition of artifacts was in practice often connected to imperial conquest. It is difficult to claim that a fossil fish belongs to any human society, but the accessibility of many geological sites had to do with empire. The situation in archaeology is far more complex: many artifacts were only discovered because of the total disregard for local peoples and customs, were removed from their sites because of assumptions of cultural superiority, and were transported to England where they were reinterpreted in their new context and with little to no thought of the living heirs of the cultures that crafted the vaunted remains. It is useful to consider two historians of anthropology who offer descriptions of the role of time in encounters with so-called primitive peoples. These theories of encounter can be usefully extended into the realms of archaeology and geology. To put it simply, Victorian observers imagined they were engaging with human traces when they encountered people who they believed to be less civilized than themselves. Anne McClintock describes the space of encounter as "anachronistic space," and Johannes Fabian argues that anthropology depends on the assumption, contrary to the prejudiced view of the Victorians, that cultures are coeval—that they exist in the same time. I would like to explore both anachronistic space and coeval encounter in application to nineteenth-century geology and archaeology.

In *Imperial Leather*, McClintock considers temporality a central element of colonial encounter, as she examines the racism and sexism inherent in imperial conquest.[11] She introduces anachronistic space to describe the supposed existence of primitive peoples outside the linear time of the conquering empire. McClintock explains:

> The colonial journey into the virgin interior reveals a contradiction, for the journey is figured as proceeding forward in geographical space but backward in historical time, to what is figured as a prehistoric zone of racial and gender difference. One witnesses here a recurrent feature of colonial discourse. Since indigenous peoples are not supposed to be spatially there—for the lands are "empty"—they are symbolically displaced onto what I call anachronistic space. . . . According to this trope, colonized people . . . do not inhabit history proper but exist in a permanently anterior time within the geographic space of the modern empire as anachronistic humans. . . .(30)

Geological and archaeological encounters were less complicated than anthropological ones. Sites of excavation unequivocally exposed a "permanently anterior time." Geological expeditions seemed to transport the observer to a distant epoch, and writers sometimes deployed the discourse of empire to make sense of the experience. For example, Lyell attempts to explain the expanse of time by likening the conquest of Egyptian tombs to the conquest of geological strata. When Victorians visited Pompeii, they insisted they had journeyed to the past, and, in London, the city's past asserts its presence and makes even the modern, forward-moving city a site of anachronism as imperial Rome rises to the surface of imperial Britian.

McClintock goes on to consider the deliberate construction of anachronistic spaces of display, notably at the Great Exhibition of 1851, and observes, "in the compulsion to collect and reproduce history whole, time—just when it appears most historical—stops in its tracks. In images of panoptical time, history appears static, fixed, covered in dust" (40). In *The Birth of the Museum*,

Tony Bennett offers a richer reading of the museum's relation to time than that suggested by McClintock. Bennett argues that museums served the vital interpretive function of reverting time back into space: "If the essential methodological innovations in nineteenth-century geology, biology and anthropology consisted in their temporalization of spatial differences, the museum's accomplishment was to convert this temporalization into a spatial arrangement" (186). He goes on to emphasize the accessibility of time in museums: "In fact, the museum, rather than annihilating time, compresses it so as [to] make it both visible and performable" (186). Bennett would take issue, I believe, with McClintock's assertion that time on display is static. Rather, exhibition provides a space for presenting an expanse of time in miniature. Baudrillard addresses such a notion of reining in time when he writes of "the time of the collection as diverging from real time." He explains, *the setting up of a collection itself displaces real time*. Doubtless this is the fundamental project of all collecting—to translate real time into the dimensions of a system" (Baudrillard's emphasis; 16). The system of collection, whether private or public, becomes a way of manipulating chronology. And collecting is not the only metaphoric vehicle for the management of time: Beer writes of narrative, "The narrative time of the *Origin* [*of Species*] is not one that begins at the beginning but rather in the moment of observation" (1983, 59). Like collection, text offers a space where time is miniaturized into a controllable system. Thus, I'd like to complicate McClintock's anachronistic space: it is not the place where time stops; instead, it is where the larger, temporal narrative is condensed into a pseudochronism— man-made and made for men to observe, narrate, and, above all, control.

Johannes Fabian recommends "coevalness" as the appropriate temporal model for encounters with the human other, and I will expand this term to describe the nature of relationships across time. In *Time and the Other*, Fabian, like McClintock, explores the function of time in an imperial context. He notes that in the nineteenth century time shifted from being "the vehicle of a continuous, meaningful story" to instead being "a way to order an essentially discontinuous and fragmentary geological and paleontological record" (14), a system like Bennett's and Baudrillard's. With this new ordering function of time arises what Fabian calls an "allochronic discourse" in which the subject observer exists in a different time than his object. Fabian here echoes McClintock, and both of them quote Joseph-Marie Degérando, who wrote in 1800: "The philosophical traveler, sailing to the ends of the Earth, is in fact traveling in time; he is exploring the past; every step he makes is the passage of an age" (qtd. in Fabian 7).[12] Like McClintock, Fabian challenges this imperial prejudice against people who not only inhabit different places but seem to inhabit different times as well. He argues that an anthropologist can only perform field work successfully if he assumes that he and his object are coeval, that they exist in the same time. He explains that many anthropologists deny coevalness: they "place the referent of anthropology in a Time other than the present of the producer of anthropological discourse" (31).

I want to complicate Fabian's definition of coevalness. At an excavatory site where the culture or epoch under examination is, in fact, from the past, it would seem there is no place for a coeval approach; the encounter is by definition allochronic. While it is racist for a Victorian traveler to consider that he has journeyed to the past when he visits an African village, it is simply accurate for that same traveler to claim he faces the past when he stands at Pompeii. Like Degérando passing through ages, the archaeologist or geologist literally explores the past as a physical space. Thus, I want to extend Fabian's coevalness of cultures to time: in the literature of nineteenth-century science, there is a modified sense of coevalness of distinct time periods—assorted pasts, the present, and the future coexist at the single point of an excavation. Yet, already I must qualify this claim. The archaeological or geological site is a trace or a collection of traces, and the trace signifies not only the past but also the passage of time. Past and present are, of course, not coexistent, and the trace reveals the temporal gap. Yet, examining a trace, the observer brings her present into direct contact with the past, and in so doing she diminishes the expanse of time or, at least, conquers it by standing outside it. The trace insists that the past is passed, yet it endures. Perhaps the Victorians longed for a coeval relationship to the past, most of all because such a relationship would implicitly require a coeval relationship also to the future. The object from the past that exists in the present and implies the future is the key to the effort to control time: the object might be a trace and thus embodies the passage of time yet it itself exists across time and paradoxically becomes a vehicle for asserting a coeval view of time. At the site of the artifact or the excavation—anachronistic space—all time is laid out together, and the depth of time collapses beneath the observer's gaze.[13]

Returning to Ricoeur, it is important to stress that the trace is a function in the mathematical sense. It is never a static object, but rather an artifact from the past, altered by time, and defined in the context of the present. It embodies multi-temporality as well as inter-temporality and becomes a foundation for the construction of anachronistic spaces and for the assumption of coevalness vis-à-vis periods of time. The trace also functions as a signifier of the imperial ideology I have been discussing. As an object from the past, it enables the nineteenth-century collector to possess, define, and subdue past time (recall Barthes describing Michelet's domestication of Death), and as an object in the present, it forces the observer to redefine himself in relation to the ever-expanding time scale and to imagine his own end as a similar artifact. The imperial context that underlies the theoretical contributions of McClintock and Fabian is more complex than I have thus far allowed.[14] The dark side of imperial arrogance is the fear that degeneration and imperial collapse lay ahead for Britian.[15] As he proudly distanced himself from supposed primitives, the nineteenth-century imperialist, however much he may have asserted his own superiority, suspected that the so-called savage was his kin. Fears of degeneration arose out of racist anxieties about contamination from already degenerate people (at home and abroad), and these fears were confirmed and deep-ened by Darwin's writings at mid-century. However, I want to argue that de-generation was at issue early in the century, long before Darwin, because of the interest in ruins and the recognition that they may picture the future as well as the past. Michael S. Roth, writing in *Irresistible Decay: Ruins Reclaimed*, states the case clearly: "The disappearance, the threat of loss, is key to the at-traction of ruins—and to their essential ambiguity. . . . Their decay is irre-sistible to us because it allows us to perceive the passage of time as irresistible. Yet ruins do resist because they persist, but not too well" (2). Faced with geo-logical and archaeological ruin, nineteenth-century observers felt they wit-nessed at once the decay of the past and a preview of their own eventual ruin, yet paradoxically they also saw the persistence of the past, and therein lay hope for the future. Such is the powerful multivalence of the trace.

The Eye of the Trilobite

Among the many traces that captured the nineteenth-century imagination, none collapsed the distance between past and present so adamantly as the trilobite. A small crustacean, the trilobite thrived in the Paleozoic era, and its fossilized remains were found in abundance during the early years of geology. Figure 1.1 shows a sketch of a trilobite from Gideon Mantell's *Pictorial Atlas of Fossil Remains* (1850). A relic from such a distant period, the trilobite chal-lenged its excavators to peer into the abyss of time stretching back beyond the scope of imagination. It evoked the breadth of Earth history, as contemporary science writer Richard Fortey muses in his enthusiastically titled *Trilobite!*: during the 300 million years they inhabited the seas, trilobites witnessed "con-tinents move, mountain chains elevated and eroded to their granite cores, they have survived ice ages and massive volcanic eruptions" (24). Fortey is de-lighted by the trilobites' longevity, but the presence of this fossilized crus-tacean was more complicated for the Victorians. In anachronistic encounters, they expressed awe in the face of such a resilient creature, but the trilobite also inspired an anxiety—if this armored, long-lived species ultimately succumbed to the passage of time and mortality, then what of the people who examine the trilobite in the present? Tennyson articulated the concern clearly in *In Memo-riam*, when he asked "shall [man] / . . . . / be blown about the desert dust, / or sealed within the iron hills?" He goes on to despair, articulating the fear of his generation that "life [is] as futile, then, as frail" (56: 8, 19–20, 25). The trilo-bite and all its fossilized kin brought individual mortality and, worse, extinc-tion before the Victorian public, who found itself unable to look away from these alarming traces.

In her comic poem "The Lay of the Trilobite" (1887), May Kendall (1861–1931) imagines a conversation between a poet wandering about "a mountain's giddy height" (1) and "A native of Silurian seas, / An ancient

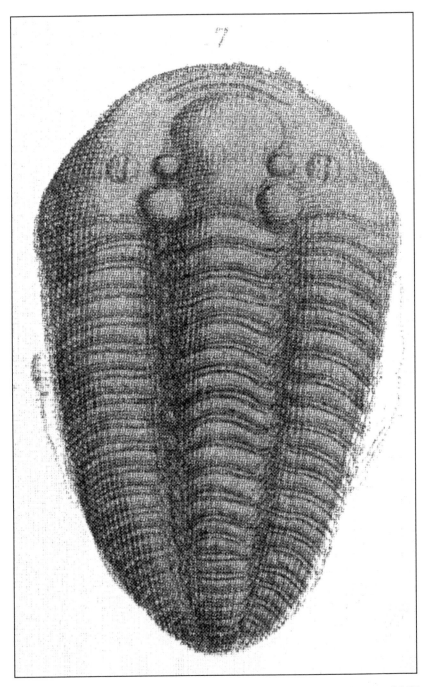

Fig. 1.1 Figure of a trilobite, Gideon Mantell's *Pictorial Atlas of Fossil Remains* (1850).

Trilobite" (7–8). The trilobite, who seems remarkably well-read,[16] references Huxley and with parallel structure makes explicit the connection between science and the loss of faith that so shook the nineteenth century. He observes, "How all your faiths are ghosts and dreams / How in the silent sea / Your ancestors were Mono-tremes" (24–26). The trilobite expresses an ignorance-is-bliss philosophy and distinguishes himself from humans in having been quite content with his simple life as a crustacean. The poet is persuaded and regrets that human life is so uncomfortably complex. Kendall concludes with the poet's lament:

> I wish our brains were not so good,
>   I wish our skulls were thicker,
> I wish that Evolution could
>   Have stopped a little quicker;
> For oh, it was a happy plight,
>   Of liberty and ease,
> To be a simple Trilobite
>   In the Silurian seas!
>
>   (65–72)

"The Lay of the Trilobite" reflects the popularity of this particular fossil type, but, despite its comic tone, it also expresses humans' regrettable ability to understand the fossil evidence before them and recognize that long-held narratives of Earth and human history are but "ghosts and dreams." With the loss of these comforting histories, the degeneration the poet longs for in line 67—"I wish Evolution had stopped a little quicker"—becomes itself a reality rather than a dream. The trilobite reveals at once how far humanity has advanced but also the ruin that likely lies ahead. In his coeval meeting with the trilobite, Kendall's poet has the opportunity to glimpse both past and future. Many surely shared the view expressed here: we would be far more content if we had remained ignorant of the geological record and all that it signifies. Kendall's poet learns this lesson, literally and figuratively, from a trilobite.

Thomas Hardy (1840–1928), always attracted to the intrusion of relics into present life and moments of contact between past and present,[17] confronts the imperiled Henry Knight with the cold stare of a trilobite in *A Pair of Blue Eyes* (1872–1873), and discards the notion of past and present separated by so many millennia stretched out in a line. Awaiting either rescue or death, Henry clings to a cliff, faced with the inevitability of his own mortality and the prospect of the mortality of an entire species:

> By one of those familiar conjunctions of things wherewith the inani-
> mate world baits the mind of man when he pauses in moments of sus-
> pense, opposite Knight's eyes was an imbedded fossil, standing forth
> in low relief from the rock. It was a creature with eyes. The eyes, dead
> and turned to stone, were even now regarding him. It was one of the

early crustaceans called Trilobites. Separated by millions of years in their lives, Knight and this underling seemed to have met in their place of death. (209)

Hardy creates a moment in which time, like Henry, is suspended, but such close scrutiny of the fossil reveals that the present, again, like Henry, does not wait alone: "Time closed up like a fan before him. He saw himself at one extremity of the years, face to face with the beginning and all the intermediate centuries simultaneously" (209). Henry and the trilobite are not alone in their coeval encounter: when one looks into the eye of the trilobite, all of time collapses together. Hardy thus expresses the conflicting views of time the trilobite and other traces inspired: the individual recognizes his own insignificance in the evident mortality of the fossilized creature, yet, in his ability to perceive the trace and transcend the expanse of time to find common ground with the fossil,[18] the individual makes himself temporally significant. In a similar scene in Tennyson's *The Princess*, the subject of chapter 3, Ida faces women's bleak future as she excavates a fossilized beast from the distant past.

While the trilobite and other fossils invited writers like Hardy and Tennyson to reimagine time, at a more concrete level, they secured acceptance of uniformitarianism. The trace of impressive geological change, the trilobite recalls the past but also becomes an object of significance in the present. The creature's body with its armor-like shell is easily spotted, but it's the eye of the trilobite that entranced those with a proclivity to think on time. The often well-preserved intricacy of this fossil-eye offers a rare glimpse of the prehistoric world, and many observers were struck by the apparent likeness of that world to their own. With a variety of lenses, the eye of the trilobite is identical in construction to the eyes of contemporary crustaceans and insects. This kinship across millions of years moved many observers who looked into the eye of an extinct Paleozoic beast and saw not distance, but, despite the eons separating them, a profound closeness. In the *Ninth Bridgewater Treatise*, William Buckland uses the trilobite to defend the uniformitarian position.[19] The eye of the trilobite suggests that processes in the past are no different from those of the present. Buckland explains that as the shape and configuration of the many lenses of the trilobite's eye are identical to those of contemporary crustaceans, "The mutual relations of light to the eye, and of the eye to light, were, therefore, the same at the time when crustaceans first existed in the bottom of the Silurian seas, as at the present moment" (qtd. in Mantell 1848, 794). Like Henry Knight, Buckland looked into the eye of the trilobite and saw the present.

The trilobite and the insight it offered into uniformity of processes was one of the most important traces forming the foundation of the science of geology. Indeed, in the eye of the trilobite, uniformity glared out at the nineteenth century, and on the stony platform of its telling gaze a paradigm shifted. In *The Structure of Scientific Revolutions*, Thomas Kuhn describes a process by which a culture comes to recognize that its way of understanding the world cannot account for certain data. When first confronting this anomaly, most people endeavor to make sense of the problem through creative application of existing principles. Eventually, some scientists break with tradition and assert the falsity of the old paradigm. Then begins the lengthy process of discovering a new paradigm, a process that involves debate at increasingly minute levels of detail. In due course, the old paradigm fades away and no longer challenges the new; at this point, conflict exists only within the new paradigm as its parameters are established. Finally, scientists and students accept the new paradigm as the norm, asserted not by specialists and theorists but by the voice of the educated mainstream. Kuhn offers several examples of such scientific revolutions; one is the transition to uniformitarian geology (47–48).

I place so much emphasis on the shift to uniformitarianism because this geological principle became a foundational strategy of asserting individual significance. It is also a scientific theory that relies not only on evidence but most impressively on imagination. In her analysis of Darwin's use of metaphor and analogy, Beer writes: he "must give primacy to imagination, to the perception of analogies, and must extend the study of forms fixed in the present moment into a study of their mutability and transience as well as their powers of transformation and of generation" (1983, 91). More simply, the observer imagines the past through the vehicle of present objects. Uniformitarianism enabled the broad interpretation of fossils and rocks that took them out of the static realm of the artifact and into the dynamic interpretive realm of the trace. It was uniformitarianism that made the trilobite's gaze, to such as Kendall and Hardy, familiar rather than utterly alien. An assumption of temporal coevalness, uniformity insists that past and present coexist, and it takes as proof the material objects uncovered and studied through excavation. Nineteenth-century excavators are both observers of small actions and actors themselves; individuals, thus, asserted their own significance by emphasizing their empowered vantage point and by insisting on the impact of minute actions. When applied to human experience, uniformitarianism offered an implied significance that became a great balm in the face of Tennyson's "Time, a maniac scattering dust" (1850, 50: 7). If the minute actions of a water drop could carve a canyon, then the seemingly insignificant actions of a man or woman might also have great effect. Uniformitarianism insists on an expanded time scale which inevitably made humans feel very small, yet it also empowers people to imagine the tremendous potency of the small.

"All Relics Here Together"

Geological excavation inevitably involved incidental archaeology. As day-geologists dig in search of a trilobite, they might uncover instead shards of roman pottery. In fact, incidental archaeology posed one of the great challenges to the traditional sense of human history: the discovery of human remains alongside the bones of prehistoric beasts like mammoths asserted the depth of human time. Some theologians countenanced geology only by believing that the geological narrative predated the appearance of humans and thus did not necessarily contradict the creation story as presented in Genesis, but the undeniable presence of human bones in the midst of geological relics made such a reconciliation of Genesis and geology difficult to uphold. Furthermore, the integration of human remains into the geological landscape made the application of geological principles to human experience impossible to ignore. If the mammoth died and his kin are extinct, then the human lying by his side and his descendents seem destined to the same fate. Thus, archaeology took up the concerns of geology: traces of past life simultaneously reveal the distance and closeness of the past to contemporary experience, and in this temporal flexibility, archaeological traces empower observers to assert the significance of not just human life, but also culture and nation and even particular, individual experience.

For example, a Roman bracelet discovered in the mudflats of the Thames poignantly suggests the mortality of the woman who once wore the jewel; it notably encircles emptiness where it once encircled flesh. Yet, at the same time, it offers a contemporary observer the opportunity to reanimate the woman as her life is rediscovered and revalued through its remains. The bracelet functions as a trace: it is an object from the past, and it signifies absence yet it asserts its presence as it becomes an object of study and interest, both scientific and aesthetic. This particular trace has an additional symbolic resonance as it inevitably directs the observer to consider the bracelet on her own wrist. As a nineteenth-century woman admires a collection of Roman remains displayed in a private museum, she cannot help but imagine the future life of her own jewels: they will signify her absence and evident mortality but they will also secure her an immortality by association.

In "The Burden of Nineveh," Dante Gabriel Rossetti explores the cultural, rather than individual, implications of archaeology.[20] Like "Lay of the Trilobite," this poem examines the moment of contact between an individual and a relic from a distant time. In both instances, the relic overwhelms the individual and reminds him of his insignificance. Rossetti's poem, first published in 1856, describes the poet's experience of exiting the British Museum and happening to witness the arrival of the great winged bull, excavated from Nimroud and shipped back to London by Austen Henry Layard.[21] From the first moment Rossetti's speaker lays eyes on the statue, he conflates the English present with the Assyrian past. He leaves the gallery that houses the Elgin marbles:

Sighing I turned at last to win
Once more the London dirt and din;
And as I made the swing-door spin
And issued, they were hoisting in
A winged beast from Nineveh.

(6–10)

The London setting is clear, and the trappings of mid-century, modern life are inescapable; the speaker pushes open the revolving door and finds himself immediately amidst London's "dirt and din," the miasmic environment that then also becomes the new setting for the Assyrian bull god. Rossetti uses the particular image of the revolving door to locate the statue in the immediate present but also to represent the movement of the past into the present and even the interchange of times. He pursues the same conflation when he marvels, "On London stones our sun anew / The beast's recovered shadow threw" (41–42). There is at once something peculiar and exciting about the shadow cast on London pavement by this beast from a distant time and place.

Indeed, the bull's shadow becomes, for Rossetti, a sort of Ricoeurian trace. He muses on the aspirations of the long-dead priests who worshipped the bull god: ". . . could all the priests have shown / Such proof to make thy godhead known? / From their dead Past thou liv'st alone; And still thy shadow is thine own" (46–49). Rossetti contrasts the "dead Past" with the living statue, its shadow being one sign of life. Moreover, while the statue was created long ago, the shadow is a thing of the present, and this continued life interests Rossetti above all else. He extends his attention to the shadow in the following stanzas as he imagines the original worshippers of the bull-god—"Within thy shadow . . . Sennacherib has knelt" (61–62)—and then, a few lines later, the Christian worshippers who also prayed in the bull's shadow: ". . . till 'neath thy shade / Within his trenches newly made / Last year the Christian knelt and pray'd" (67–69).[22] He goes on to imagine the subdued shadow that the bull will cast once installed in the museum gallery, where "blank windows blind the wall" (72): shadows are cast not by the sun but by artificial light, and worshippers are replaced by school children. Rossetti uses the shadow to imagine the bull's history and its immediate fate, but he also uses the shadow to reflect the bull's continued life outside its original historical moment. Like a fossil, the bull is a material object crafted long ago, yet it has a life in the present, a life of renewed interpretation and significance in the pseudochronic space of a museum.

Rossetti emphasizes the significance of the winged bull for the faith of his Victorian contemporaries. Anticipating Baudrillard, he reflects on the museum's power to annihilate historical process as it blends together the relics of different ages. Andrew Stauffer describes Rossetti's response to the museum in "Dante Gabriel Rossetti and the Burdens of Nineveh": "In a very literal sense, then, Rossetti's response to the colossal Assyrian bull grows out of the basic contradiction of the museum itself, which offers each appropriated relic to be understood even as it confounds them all in a physical proximity that occludes temporal and geographical distances" (389). Rossetti mentions the mummies housed a floor above the bull (101–2) and remarks, ". . . Nay, but were not some / Of these thine own 'antiquity'? / And now,—they and their gods and thou / All relics here together" (104–7).[23] The significance of each god is diminished as it is displayed alongside other gods, and the extension to the relics of contemporary Christian religion is implied: if the sacred objects of ancient cultures have lost their religious value and have become merely artifacts, then are not the signs of contemporary religion doomed to the same fate? Rossetti imagines a future time when the bull's origins have grown obscure and "some may question which was first, / Of London or of Nineveh" (169–70). His fantastic future archaeologists will not be able to discern whence the statue originated and will interpret it as "a relic now / Of London, not of Nineveh!" (179–80). He extends the speculation into a direct challenge to history:

> Who, finding in this desert place
> This form, shall hold us from some race
> That walked not in Christ's lowly ways,
> But bowed its pride and vowed its praise
> Unto the god of Nineveh.
>   (186–90)

He calls into question the archaeologist's power to accurately interpret his or her finds, but he also marvels at the object's power to transcend time and context and take on a significance of its own. As a trace from the past but with a life in the present and presumably also the future, the winged bull became for Rossetti a sign of the power of the material artifact. It recalls history, but, more, it insists upon reinterpretation of the present. "The Burden of Nineveh" describes the burden of London as it struggles to assimilate the past into its aggressively progress-minded present. Stauffer asserts that Rossetti "appropriates the alien artifact as a figure for imperial hubris" (379). What will be the future of London? Rossetti emphasizes the point: those who observe the bull-god in the British Museum will suffer the same fate of the god's worshippers. We, too, will be forgotten. Yet, even Rossetti does not offer such an unrelentingly bleak view of the human future. Remember the revolving door with which Rossetti began: that statue and the Victorian observer spin round, entering and exiting the museum in turns. The distance between them is diminished and, like the Egyptian and the Assyrian relics, they are "all relics here together." Thus, the school child studying the bull is doomed to extinction but she may yet cast her shadow on some future city. With the ravages of time's passage comes also lasting preservation.

Excavating Victorians

Excavating Victorians contributes to the fast-growing body of work in Victorian cultural studies with its use of a methodology employed also by the Victorians and thus uniquely suited to the study of their literature. A term that I use both as literal descriptor and metaphor, "excavation" also names a particularly Victorian hermeneutic: it demands the simultaneous interpretation of an object and of its situation, and it focuses on the recovery of the past while insisting on the primary act of interpretation in the present. Progress and decay, both compelling tropes for the Victorians and us who study them, meet in the act of excavation. Indeed, digging down into the layers of the past inevitably reveals narratives that press into the future. To negotiate the complexities of this temporal instability, I follow the model of excavation, considering the context in which I find my text-objects but keeping my attention closely focused on the objects themselves.

In my application of literary analysis to the excavatory sciences and the works of poetry and fiction written with those sciences in mind, I have determined that the nineteenth century did far more than discover time; pivoting on the potency of the trace, science and art shaped a path for the individual through time's roar. Though slowed "by the eternal note of sadness," the nineteenth century figured out how to "begin again," how to discern the rattle from the "Ah!" and comprehend their harmony. Examination of traces positions the observer at a distance from time: he inhabits an anachronistic space where the oppression of the vast temporal narrative is held at bay, and he is empowered to comprehend all of time and to narrate its history. Central to my study is the notion of text. The scientist explicitly reads artifacts, and then translates that text into a narrative of his or her own. It is not my aim to report the history of geology and archaeology or to philosophize science. Rather, I aim to excavate Victorian culture through close analysis of the literature and the literary themes of excavation. As the Victorians traced time through its vestiges, so do I trace their efforts to survive time's expansion through their narrative assertions of individual worth.

The first section of the book focuses on geology, the new science of fossil traces. Examining trilobites and "rude bones," as Tennyson's Ida labels geological evidence in *The Princess*, led Victorians to reconsider their own position in the world. They felt at once diminished by the vastness of the time scale and empowered by their ability to interpret it in its traces. Writers and scientists assembled medleys of fossil remains and human artifacts to make sense of the past, the present, and of time itself. Chapter 2 analyzes writings by early geologists Charles Lyell and Gideon Mantell that emphasize the authority of the geologist not only as a scientist but also as a writer. Historians of science have discussed the contributions of these men to the advance of geology in the nineteenth century; I attend instead to their narrative strategies and to their rhetoric, and I argue that the observer and the interpreter are central to the stories Lyell and Mantell tell. This analysis complements other studies that stress the doubt induced by science in the nineteenth century by showing that these geologists maintained faith in the significance of the individual.

In my discussion of geology, I emphasize the act of interpretation and the narrative product that results from geological excavation. Closing the gap between past and present at the site of the trace and in the interpretive text lends authority to the writer, who thus seems to control time. In chapter 3, I extend my consideration of efforts to maintain control in the face of the expanded time scale with an analysis of Tennyson's *The Princess* (1847–1850) as a fairy tale wherein "nature red in tooth and claw" is domesticated. *The Princess* is subtitled "a medley" and can be read as a collection of discrete literary objects: the narrative poem broken by the interpolated songs, the tale of the past vying with voices of the present. This form reflects the new way of reading the past through so many fossils, bones, and artifacts laid out in collections and texts. The form also connects the poem to comedy and to music, and I argue that its generic confusion reflects the struggle to achieve narrative authority at a time when agency has been replaced by process. The heroine Ida voices the despair of her generation: she strives to revolutionize the role of women but is frustrated in her attempt as she imagines a future as nothing more than "rude bones." Yet, in *Excavating Victorians* I read the poem's comic resolution as an affirmation of the value of human life and love despite the abyss of time.

The second section of the book turns to archaeology and the human dimension of a history derived from material remains. Chapter 4 explores the literature describing excavations at Pompeii and London. Of interest for both its tragic ruin and its extraordinary preservation, Pompeii was at once "The City of the Dead" and a city still living. Writers recognized London in Pompeii and imagined a future London similarly preserved. Past and present were most directly conflated when Londoners saw Pompeii in contemporary London, where work to improve the urban landscape often made the city appear as a site of archaeological excavation. Archaeology, like geology, reveals worrying evidence of the loss of life and civilization, yet it also offers the possibility of evading total annihilation by leaving behind lasting traces. The peculiar conjunction of preservation and destruction along with the power of the writer to craft a narrative whole out of archaeological traces offered hope to readers worried about the prospect of life and civilization in ruins.

Chapter 5 considers Dickens's broad use of archaeology and excavation in *Little Dorrit* (1855) and *Our Mutual Friend* (1864). Archaeological imagery underscores Dickens's persistent interest in urban progress and decay. In this, he follows his contemporaries, discussed in chapter 4, who make connections between works to improve the city and its being rendered an archaeological ruin. He joins with geologists and archaeologists to extol the power of the reader and writer to manipulate the traces of the past and thus to move steadfastly into the future. Like Tennyson, he also echoes Michelet with his attention to collectors and to those who assemble fragments. In such articulation lies the possibility of coherence when the scientific evidence suggests only fragmentation and purposelessness. Though Dickens certainly offers a bleak view of life in Victorian England, he ultimately sees hope for those who read and interpret traces and who attend to making sense.

As I move throughout the book and within chapters from nonfiction prose to fiction and from science writing to poetry, my reader will see that this body of literature shares the common aim of facing the past, a past newly understood to be vast beyond imagining, and rescuing the individual from being lost in the abyss. The texts considered here also share an emphasis on the authority of the writer, who is inevitably also an interpreter, and on the temporal conflation that occurs and empowers when objects (whether voices, fossils, or quotidian debris) are assembled into a narrative whole. Tennyson's medley, Dickens's mystery, Lyell's attention to the observer, Mantell's emphasis on the collector, and the many who played at reanimating the past and controlling the future in Pompeii and London together form a medley, a collection of my own. In excavating the traces assembled here, my reader will discover a Victorian period fraught with anxiety yet hopeful, a people keenly aware of the remoteness of the past yet striving to "close up time like a fan," and, above all, a body of literature that celebrates the individual life and the power of the individual, despite the depth of time and, in some cases, because of it, to affect great and lasting change.

2

The Victorian Geologist
Reading Remains and Writing Time

Amidst the din of rushing waters, the noise from the stones, as they rattled one over another, was most distinctly audible even from a distance. This rattling noise, night and day, may be heard along the whole course of the torrent. The sound spoke eloquently to the geologist; the thousands and thousands of stones, which, striking against each other, made the one dull uniform sound, were all hurrying in one direction. It was like thinking on time.

—Charles Darwin, *The Voyage of the Beagle*

The rattle of rocks and water sounded throughout the nineteenth century, and "Time! Time! Time!" echoed everywhere, rattling the proponents of extant paradigms.[1] Geology first heard the noise, but it was not long before nearly everyone found him- or herself living amidst the din. Well into the nineteenth century, those discomfited by the mounting scientific evidence looked to the Bible for time's story, but gradually the six-thousand–year chronology recounted there was drowned out by the persistent rattle of uniform change that revealed an alarming depth of time. Compelled to accept the principle of gradualism, the nineteenth century revised time, and geologists broadcast the paradigm shift. In addition to extending the time scale, they offered the erosive action of water and other natural processes as the Earth's agents of change and thus shattered traditional notions of authorship. No observer could witness gradual geological changes in the real time of millions upon millions of years, but the geologist reads the record of such change in fossils and strata, and he re-presents time's story.[2] Rocks and the science that excavates them roared loudly throughout the nineteenth century, and many writers asked the pressing question, how does one live amidst the din? I examine the literary answers proposed by geologists, particularly Charles Lyell and Gideon Mantell, both popular science writers who shared the common aim of rescuing the individual from temporal insignificance.

James Hutton's infinite chronology—"no vestige of a beginning,—no prospect of an end" (304)—dwarfed biblical time and threatened the Judeo-Christian account of creation and human history. Rather than reading time as the passage from "the beginning" (Genesis 1:1) to the final revelation, Hutton and his followers eschewed the catastrophic events recounted in the Bible (i.e., Noah's flood) and instead argued for the gradual change wrought by the constancy of process over vast expanses of time, the theory of uniformitarianism, defined succinctly by Lyell in his full title *Principles of Geology, Being an Attempt to Explain the Former Changes of the Earth's Surface, by Reference to Causes Now in Operation*. Introducing uniformitarianism at the beginning of the nineteenth century, John Playfair wrote in *Illustrations of the Huttonian Theory of the Earth* (1802), "The rivers and the rocks, the seas and the continents have been changed in all their parts; but the laws which direct those changes, and the rules to which they are subject, have remained invariably the same" (qtd. in Lyell, *Principles* vol.1, no page). In *The Abyss of Time*, Claude Albritton explains that uniformity is the "geological corollary of the logical principle of simplicity (Occam's razor)" (215).[3] Absent other evidence, it is logical to assume the constancy of processes. Such an assumption has the extraordinarily important consequence of empowering the geologist to study a past he could never see. Uniformitarianism thus depends not only on a broad expanse of past time and the theory that actions observable in the present are analogous to those of the past, but it depends also on the geologist himself who performs the fundamental scientific act of observation and the fundamental narrative act of storytelling.

As described in chapter 1, this emphasis on the scientist is central to uniformitarianism and to my broader reading of the literature of nineteenth-century geology. In this chapter, I will demonstrate that the geologist assumes the authority of the storyteller of the Earth's history, and in this role, he asserts the continued significance of the individual despite the abundant scientific evidence that suggests otherwise. Writing of Darwin's professed proclivity for storytelling in his youth, Gillian Beer concludes, "his passion for fabulation expressed both a desire for power and an attempt to control the paradoxes by which he was surrounded" (1983, 25). Describing fossilist Mary Anning, Judith Pascoe makes a similar point: "The poetic gleanings that are preserved in Anning's commonplace book draw our attention to the literary aspects of early-nineteenth-century natural history writing, the ways in which poetry was enlisted to help make sense of inscrutable and terrifying fossil remains" (142).

The inviting conjunction of story and control is central to the work of the uniformitarians; indeed, the scientific theory of uniformity depends on narrative, without which the testimony of the rocks would not be witnessed. In *Great Geological Controversies*, A. Hallam observes that uniformitarianism describes both a system and a method: the system of gradual and constant actions that occur over countless millennia, and the method of observing those actions in the present and thus also observing the past by proxy. He distinguishes uniformitarianism in this respect from catastrophism, which is a system only and does not make the geologist central to the narrative. Catastrophism interpreted the evidence of the geological record to reveal a series of great catastrophes, most notably the flood, and thus managed to integrate geology into the biblical creation story. Catastrophism complemented natural theology, which remained a strong and vital scientific movement throughout the nineteenth century.[4] Most important for my purposes, catastrophism (as it was most often presented in the nineteenth century) deferred to the traditional narratives of natural history.

Not all geologists readily relinquished the Mosaic cosmogony, the biblical Earth history that sets the creation at approximately 4000 BCE and takes the flood as its central event.[5] Indeed, early debates within the discipline centered on the nature and veracity of the biblical flood. William Buckland (1784–1856) and Robert Jameson (1774–1854),[6] for example, saw traces of Noah's flood waters in the ammonites embedded in mountaintops—fossils and strata offered illustrations of God's work. In a seminal text on this subject, Charles Coulston Gillispie describes the conjunction of Genesis and geology as a struggle to locate "religion (in a crude sense) *in* science rather than one of religion *versus* science" (Gillispie's emphasis; ix). Indeed, geologists and theologians alike sought to forge a seamless bond between the Bible and the new Earth science. Sandra Herbert notes that "diluvium," debris from the floods, was a key term in the lexicon of early geology (72). Such integration of religion and science appeased the Mosaic cosmogonists, certainly, but it also fulfilled the professional obligations of many university-connected geologists. In his biography of Buckland, Nicolaas Rupke claims that the professor had to espouse a Mosaic view to placate the Anglican authorities at Oxford. He writes, "the diluvial theory became the linchpin by which modern geology attached itself to the carriage of the Anglican tradition and the clerical purpose of an Oxford education" (60). Rupke suggests a more complicated relationship between Buckland and the traditional Christian doctrine than has generally been posited, most notably by Gillispie. For instance, Rupke explains that Buckland considered that the history described in Genesis was meant to be only the history of the human experience on Earth: millions and millions of years and countless geological changes occurred before the biblical narrative begins.

As Buckland demonstrates, the scientific evidence did not require a move toward atheism, yet for many such a move became inevitable. For many, the gradualism Hutton and later Lyell described quietly denied the necessity of God. In 1851, John Ruskin (1819–1900) expressed the frustration of his generation, the first to pass its adult life under the shadow of geological evidence: "[My faith] . . . is being beaten into mere gold leaf, and flutters in weak rags from the letter of its old forms. . . . If only the Geologists would let me alone, I could do very well, but those dreadful hammers! I hear the clink of them at the end of every cadence of the Bible verses."[7] Increasingly, scripture and science seemed to be at odds. Huge canyons were not formed by a single cataclysmic event; rather, they were carved by the persistent actions of small streams or even water drops working steadily epoch after epoch ad infinitum. Loren Eiseley writes of Hutton: "So preternaturally acute was his sense of time that he could foretell in a running stream the final doom of a continent" (73). Hutton may have not followed such a stream to its atheistical end, but for many the evidence revealed through uniformitarian study foretold also the doom of the biblical creation narrative. Nature self-constructed—an epic without beginning, end, or author—challenged the traditional narrative of Earth history. Uniform change does not require a creator; nor does it even require an explanatory text—the rocks, with the geologists' help, tell their own story. Emphasizing the authority of the scientific observer, many uniformitarian geologists reluctantly expunged the author of creation.

In this chapter, I focus on two geologists who successfully disassociated Genesis from geology and who worked to craft from the fossil record a new story of the Earth. Lyell and Mantell ascribed the geologist with the authority to report Earth history, and they turned away from the question of divine authority.[8] Rudwick (1970) describes the astute rhetorical strategies Lyell employed to avoid entering theological debates; for instance, he argues that Lyell carefully confined his attacks to the pagan cosmogonists of the past rather than challenging the theologians of his present. Indeed, *Principles of Geology* contains no overt consideration of the Earth's origins. George Poulett Scrope (1797–1876) praised this omission in his 1830 review of *Principles* published in the *Quarterly Review*: "To bring forward the scriptures as the foundation of geology, or geological hypotheses as support to the scriptural relations, is to degrade the sacred writings, as well as to impede the progress of knowledge" (414).[9] Lyell responded to Scrope in a letter: clearly referencing Hutton's famous conclusion to his *Theory of the Earth*, he mused, "Probably there was a beginning—it is a metaphysical question, worthy a theologian—probably there will be an end . . . It is not the beginning I look for, but proofs of a progressive state of existence in the globe" (1881, 269–70). Focusing the discussion of geology on the nature and pace of change, rather than on the traditional narrative features of beginning, middle, and end, Lyell established an alternate Earth history: neither refuting nor complementing the chronology recounted in the Bible, Lyell examined rocks, fossils, and bones and found that the Earth tells its own story. The geologist who interprets and records that story wields tremendous authority.

In the absence of the familiar theological narrative, metaphors of reading and writing infused geology with a sense of interpretive control. In *Darwin's*

Plots, Gillian Beer notes that Lyell uses these metaphors of etymology to describe the Earth; she carefully scrutinizes Darwin's adaptation of these tropes of narrative and argues that he takes this particular metaphor directly from Lyell. Whether or not Lyell crafted this metaphor, it certainly resonated and became almost a cliché in the geological literature of the 1830s and 1840s. Most geologists writing in this period demonstrated keen awareness of their role as readers of time: they sought to "read the history of the past" (Mantell 1848, 657). Again and again geologists described the Earth as a complex and fragmented book that they alone could interpret. In volume 3 of *Principles of Geology*, Lyell wrote, "a study of present processes is the 'alphabet and grammar of geology'" (7). Poet and curator of the Mantellian museum, George Richardson (1796?–1848) applied textual metaphors to fossils and dinosaurs: "Wondrous shapes, and tales terrific / Told in Nature's hieroglyphic; / Written in her countless volumes" (qtd. in Rudwick 1992, 76). In *Vestiges of the Natural History of Creation* (1845), Robert Chambers (1802–1871) described a stone book and its leaves (46), and Mantell exhorted his readers to "learn to read the records of creation in a strange language" (*Medals* 26).[10] Emphasizing the difficulty of reading the Earth's imperfect record, Darwin famously wrote:

> For my part, following out Lyell's metaphor, I look at the geological record, as a history of the world imperfectly kept, and written in a changing dialect; of this history we possess the last volume alone, relating only to two or three countries. Of this volume, only here and there a short chapter has been preserved; and of each page, only here and there a few lines. (1859, 99)

From each geology text, I could extract at least one such comparison. Fossils or strata are letters of the alphabet, chapters of a book, volumes in a library, even characters in an epic. Figure 2.1 shows an image of a dinosaur—the most wonderful character in the epic of Earth history—reading, thus illustrating the extent and complexity of the metaphor of reading in early geology.[11]

Reading traces gave geologists insight into time, and more importantly, seeing themselves as readers encouraged geologists to consider Earth history as a narrative, a narrative that geologists were at liberty to reconstruct in their own texts. In *The Wonders of Geology*, Mantell describes the geologist's power:

> Before we can attain that elevation from which we may look down upon and comprehend the mysteries of the natural world, our way must be steep and toilsome, and we must learn to read the records of creation in a strange language. But when this knowledge is once acquired, it becomes a mighty instrument of thought, by which we are enabled to link together the phenomena of past and present times, and obtain domination over many parts of the natural world. . . . (26)

Fig. 2.1 "The Age of Monsters," John Cargill Brough's *The Fairy Tales of Science* (1859).

In this example of reading the Earth, Mantell daringly conjoins the "mysteries of the natural world" with "the records of creation." His description of the "strange language" in which the creation story is recorded obliquely refers to the ousting of the biblical creation and the need for interpretation of this strange, geological language. Reporting the Earth's story to the public, the geologist usurps the Bible's authority. For Mantell and other Victorian geologists, the transformation from reader to narrator resulted in "domination over . . . the natural world." Lyell makes a similar claim in his review of Scrope's *Memoir on the Geology of Central France*: "we assert [the mind's] superiority more clearly by enlarging our dominion over time . . . all the ages, even though they be myriads of years, over which science may enable us to extend our thoughts, and whence we may derive, by studying the records of the past, fresh ideas—power—intelligence—and delight" (474). Note that "studying the records of the past" is a mere parenthetical phrase, while "—power—" is the object of the sentence, arguably the object of the science. For Lyell and Mantell, reading traces of the past and representing their stories led to "dominion over time." In this emphasis on domination, we see the power ascribed to the historian by philosophers like Hegel and Michelet. It is the historian who mediates our encounter with the past, and it is the historian's present that determines the interests that the story must oblige. I think also of George Eliot's pier-glass parable: we see the patterns etched on the glass as shown by a particular light. Lyell and Mantell shine their lights, and in so doing, they essentially create the past. If history (even Earth history) is a knowledge by traces, as Ricoeur claims, than the geologist-writers are the ones who know the traces and who determine how their readers shall know them.

"A Bliss Like That of Creating"

Lyell was arguably the most important figure in nineteenth-century geology: *Principles of Geology* defined and popularized the new discipline. Stephen Jay Gould writes that *Principles* is "the book that most geologists regard as the founding document of their discipline's modern era" (101), and Martin Rudwick begins his introduction to Lyell's *Principles* with the claim that the text "is on any reckoning one of the most important books ever published in the Earth sciences" (vii). Lyell's uniformitarian approach to the geological record became the foundation not only of modern geology but also of Darwin's evolutionary biology, and even archaeology and anthropology. Mott T. Greene describes Lyell as "a kind of torchbearer in a historical relay race—having laid the foundation for modern geology, he passed inspiration on to the founder of modern biology, connecting the cycle of revolutions" (25).[12] *Principles* was immediately celebrated as a revolutionary text: Scrope wrote in the *Quarterly Review*, "we hail with the greatest satisfaction the appearance of Mr. Lyell's work, which henceforward, we can hardly doubt, will mark the beginning of a new era in geology" (417).[13] Many of Lyell's contemporaries shared Scrope's enthusiasm; even those who were not fully persuaded by Lyell's articulation of the principles of geology recognized the importance of the work. William Whewell wrote in the *British Critic*:

> Though we are not prepared to assent to all the opinions which he propounds, it is impossible to think otherwise of his work than as a most skillful and masterly attempt to combine into a consistent view a large mass of singularly curious observations and details, which no one but an accomplished geologist could have put together. (qtd. in L. Wilson 292)

Eleven editions of *Principles of Geology* were published by the time of Lyell's death in 1875,[14] and the respect Lyell won from the publication of the book secured him positions of leadership in the Geological Society and led to a prolific and always well-respected career.

Lyell changed conceptions of Earth history because he avoided religious controversy, because he was an exceptional prose writer, and because his book reached the public when they were ready to absorb the impact of uniformitarian principles. He did not invent the principle of uniformity, nor was he the first to conceive the expanse of geological time.[15] Many historians of science have charted the expansion of the time scale and the development of geology: Martin Rudwick, Claude Albritton, Paolo Rossi, and William B. N. Berry, to name just a few.[16] In seeking the origins of modern geology, Rudwick and Albritton both begin with Leonardo da Vinci, who observed, "Above the plains of Italy, where flocks of birds are flying today, fishes were once moving in large shoals" (qtd. in Albritton 20). Like that of the Earth itself, the origin of geology is difficult to find, and it is not the aim of this book to contribute to that search. In *Darwin's Century*, Loren Eisely offers a tidy and convincing history of the developments that led to Darwin's theory of evolution and led also, I argue, to Lyell's paradigm-shifting work. He highlights the contributions of James Hutton, William Smith, and Georges Cuvier, and I summarize his claims here to offer my reader a glimpse of the foundation on which Lyell built.

James Hutton first articulated the depth of time, although Eiseley acknowledges earlier men who contributed to the expanded time scale. Benoit de Maillet (1656–1738), for one, theorized that the Earth was more than two billion years old, and Georges Buffon (1707–1788) argued for a time scale in the millions of years based on his understanding of stratigraphy.[17] Eiseley and Rudwick both remind their readers that facts we now recognize as fundamental to geology—fossils are organic remains, and layers of rock represent different geological times—were novel, even radical, propositions not so long ago. By the beginning of the nineteenth century, fossils were recognized as the traces of dead flora and fauna, but geologists still had reached no consensus about how those fossils were disseminated so widely. Eiseley's second figure is William Smith (1769–1839), who, like Buffon, raised awareness that strata represent distinct epochs and can be used to gather evidence about different time periods.[18] Fossils found in the same layer of Earth can be assumed to have co-existed; fossils found above and below represent life forms that inhabited the Earth at a different time.[19] The third scientist Eiseley hightlights is Cuvier (1769–1832), whose work in comparative anatomy finally forced the scientific community to accept the troubling concept of extinction.[20] Once these three scientists had made their contributions—an expanded time scale, an understanding of strata, and the recognition of extinction—Lyell was prepared to read the rocks and report a story reliant on uniformity.

Roy Porter describes Lyell's reputation, reminding us that the first to craft the narrative of Lyell's history was Lyell himself:

> For, true or false, fair or unfair, Lyell's autobiographical vision of himself as the spiritual saviour of geology, freeing the science from the old dispensation of Moses, has exercised an unbroken fascination over almost all who have struggled to unravel the history of British geology. . . . Lyell, one can be sure, would have heartily endorsed the evaluation of himself as geology's great and successful revolutionary . . . (91)[21]

Porter goes on to describe how Lyell's history of geology included in the opening chapter of *Principles* was carefully crafted to foreground Lyell's own contributions to the developing science: "He forged myths about his own place in the geological pantheon" (91). Porter offers the compelling conclusion that in stripping away the nuance of the history of geology and representing it as a simple struggle between uniformitarians and catastrophists, Lyell himself behaves as a catastrophist: "Caricature heroes and villains loom out of the empty backdrop of history. . . . And, of course, Lyell's teleology culminates with himself making a catastrophic revolution in geology" (98). In *Time's Arrow Time's Cycle*, a history of the metaphors that drive geology, Stephen Jay Gould refines the claim that *Principles of Geology* was the "most important book every published." He says, "Lyell's *rhetorical success* must rank among the most important events in nineteenth-century geology" (my emphasis; 119). In other words, Lyell's science was not wholly original, but his writing broadcast uniformity and the depth of time in a manner compelling to a readership just ready to reimagine time.

With metaphors of reading and writing, Lyell made imaginable the revised time scale of millions of years, labeled "deep time" by twentieth-century science writer John McPhee in *Basin and Range*. McPhee accurately describes the nineteenth-century perception of time as palimpsestic, each stratum giving way to a deeper one, and each layer signifying an epoch of millions of years. Gould explains how the concept of deep time is essential to a true understanding of uniformitarianism: "The major stumbling block for this method lies in a popular perception that change so slow is no change at all; but deep time provides a matrix that converts the imperceptible to the mightily efficient" (90). Lyell's greatest challenge in *Principles of Geology* was to convey this sense of deep time: many of his readers were already familiar with and receptive to the uniformitarian methodology, but to offer gradual, barely perceptible actions as the agents of what had previously been considered catastrophic change required an expanse of time far greater than most people had yet imagined.[22] One of our greatest abstractions, time is difficult to conceive without analogy or metaphor; geology offers both. Layers of rock extend deep within the Earth, forming a vertical and visible time line. They also reveal a fragmentary narrative of Earth history, and thus each layer of rock represents an age. All the layers together represent history, and a cliff or cutting that reveals multiple strata at once is a temporal collage, an assemblage of traces. Lyell uses language in several of his writings that suggests depth as an appropriate measure of time: in his review of Scrope's *Memoir*, he refers to "all discoveries which extend indefinitely the bounds of time" (474); in chapter 4 of *Principles*, he likens Hutton's views of "the immensity of past time" to Newton's, writing "worlds are seen beyond worlds immeasurably distant from each other . . ." (1830, 63). His diction—words like "bounds," "immensity," "beyond," and "immeasurably"—reflects a physical sense of time akin to McPhee's deep time and emphasizes the crucial conjunction of space and time. Though "deep time" was not a term used in the nineteenth century, it precisely reflects the sense of time as spatial and, more specifically, as deep rather than as broad or wide. The concept of depth reinforces the fact that geology is a science based on digging into the Earth, and that the new time scale was a direct product of those excavations. Deep time thus also reinforces my own emphasis on excavation as a central act and ideology of the relationship between the individual and time in the nineteenth century.

Leaving the Mosaic cosmogonists and their six-thousand-year-old Earth in the dust, Lyell skillfully evoked the unimaginable time required to carve a canyon with a water drop. He regarded himself as a reformer, writing to Mantell in 1828 on the subject of his embryonic *Principles*, ". . . so much is to be reformed and struck out anew . . ." (1881, 177). Here, as in his formal writing, Lyell evades mention of the biblical approach to Earth history, but he implies that his project effectively overthrows that tradition. His diction conjoins the language of geology—rocks are "formed" and debris is "struck out"—with the language of religion, "reform." Although rocks and fossils were the primary tracts offered by this reformer, Lyell eases his readers into deep time with a more familiar text, an analogy to human history. He writes:

> As the present condition of nations is the result of many antecedent changes, some extremely remote and others recent, some gradual, others sudden and violent, so the state of the natural world is the

result of a long succession of events, and if we would enlarge our experience of the present economy of nature, we must investigate the effects of her operations in former epochs. (1830, 1)

Lyell invokes his readers' familiarity with interpreting history to prepare them to interpret the geological evidence of Earth history. He refers to the "present economy of nature," and then in the following independent clause, he links that economy to the "operations in former epochs." His syntax demonstrates an ambiguity designed to lure the reader into sharing his uniformitarian view. What the phrase "in former epochs" modifies is unclear: are the "effects" under observation those of the past or of the present? With this ambiguity, Lyell suggests a mutual relation between the processes of the present and those of the past no different in geology than in history.

Lyell pursues a similar logic throughout the first chapters of the first volume of *Principles*. He alludes to human history to debunk his catastrophist opponents without insulting specific individuals. In his introduction to *Principles*, Rudwick writes:

Like many would-be reformers, Lyell rewrites history in order to demonstrate that those whose present views he is attacking are inheritors of a historical tradition that has had a retarding influence on the progress of the science. He contrasts this with the more enlightened tradition in which he finds his own antecedents, though in fact he spreads his praise thinly, and thereby accentuates his own innovative originality. (xvi)

Lyell repudiates ignorant thinkers of the past and impels his readers to join him. The "Egyptian Cosmogonists" he describes in his second chapter ignorantly misread the Earth's text: "in a rude state of society, all great calamities are regarded by the people, as judgments of God on the wickedness of man" (10). His chapter title associates these historical figures with Lyell's contemporaries, the Mosaic cosmogonists. Thus Lyell implicitly accuses these contemporaries of misreading. He offers a choice between the members of the "rude state of society" and a good reader, the "celebrated" Ovid, as he quotes from *The Metamorphoses*: "'Nothing perishes in this world; but things merely vary and change their form'" (13). Ovid aptly read gradual and constant change in the world around him, and Lyell admires his assertion of a proto-uniformity. Lyell's complex analogy distinguishes usable pasts from uninformed mythologies and, without making any direct accusations, Lyell divides his contemporaries into parallel camps, the enlightened and the woefully ignorant. He invites his readers to join him in the advance of geological, rather than cosmological, principles.[23]

Lyell ends the second chapter of *Principles* with an explicit acknowledgment of the reader's authority: he suggests that his philosophy offers access to the mysteries of the Earth that eluded the pre-Christians. Lyell writes of these early thinkers:

> They had studied the movements and positions of the heavenly bodies with laborious industry, and made some progress in investigating the animal, vegetable, and mineral kingdoms; but the ancient history of the globe was to them a sealed book, and, although written in characters of the most striking and imposing kind, they were unconscious even of its existence. (20)

His reference to the written characters of geology conveys the promise that *Principles* provides instruction in reading the traces of the Earth. Lyell claimed geology would unseal the globe's history and reveal uniformity as its theme. The geologist reads the Earth-text and writes a narrative, in turn read by readers prepared to consider a revised narrative of Earth history. The geologist's authority, rooted in his ability to read well, lends similar authority to any reader—with the ability to interpret comes narrative authority.

The narrative that Lyell extricated from the Earth and promoted in *Principles* is proto-modern in its disregard of progressive, linear structure. Most historians of science associate Lyell with a cyclic model of time. Lyell rejected the progressive notion of the Earth moving ever forward in some ameliorative fashion; he famously predicted the return of a tropical climate and species akin to dinosaurs. To dramatize this prediction, the only one for which Lyell was roundly criticized, Henry De la Beche drew what became a famous cartoon of a Professor Ichthyosaur lecturing to a group of reptiles. (See figure 2.2.) Describing cycles of change, Lyell wrote to Mantell early in 1830, "All these changes are to happen in future again, and iguanodons and their congeners must as assuredly live again in the latitude of Cuckfield as they have done so" (1881, 262).[24] However, reading change as purely cyclic reduces time's complexity, and Lyell also embraced the linear. His use of a historical narrative early in *Principles* compromises his commitment to cyclic time as it illustrates the linear forces of progress and decay.[25] The shift from the darkness of the ignorant past to the enlightened thinkers of the present reveals intellectual progress, but the careful reader recognizes the catastrophists of the present in the ignorants of the distant past. Thus, the progress toward clear thinking is clouded by the decay of those who have degenerated from the wise Ovids of the past to the Paleys of the present.[26] Furthermore, for Lyell progress is decay: the future according to his theory will be a return to the past. I argue, contrary to the prevailing views, that Lyell represented change as both linear and cyclic: in his view, progress and decay are the Earth's texts, and uniformity is the principle guiding both change and the nature of time.[27] His position is in accord with the coeval one I discuss above: different epochs seem to coexist.

AWFUL CHANGES.

MAN FOUND ONLY IN A FOSSIL STATE.——REAPPEARANCE OF ICHTHYOSAURI.

J. H. T. de la Beche del & lith. 1830.

Fig. 2.2 "Awful Changes: Man found only in a fossil state—Reappearance of Ichthyosauri" by Henry Thomas De la Beche (1830), Francis T. Buckland's *Curiosities of Natural History*.

If processes in action millions of years ago are identical to those of the present, then the past seems immediate and knowable. Geologists can even imagine that they see the past with their own eyes: recall Hardy's Knight meeting the trilobite's gaze. Yet, Lyell reminds the reader that much of the geological record is merely metonymic, that fossils and rocks are traces signifying absence as much as they assert their own presence. The oldest rocks are descendents of yet older ones that no longer exist: ". . . when the thoughts had wandered through these interminable periods, no resting place was assigned in the remotest distance. The oldest rocks were represented to be of a derivative nature, the last of an antecedent series, and that perhaps one of many pre-existing worlds" (1830, 63). However old we may perceive the Earth to be, it is ever so much older. Lyell invites his readers to stretch their minds back as far as possible, and then he asks them to imagine that they have only begun to fathom the depths of time. In *Darwin's Plots*, Beer stresses the importance of imagination not only in communicating science but also in scientific interpretation: we see Lyell evoking imagination at once to communicate and to comprehend. Lyell cannot draw the image of the distant past, but he can encourage his readers to perceive that epochs lie beyond the scope of their vision. His project is not to describe the past, but through its indescribability, to gesture toward it. The readers need not know *what* deep time is; they need only know and firmly believe *that* it is. Once they believe in deep time, Lyell's readers can appreciate the temporal implications of uniformitarianism, and they can read the past in the metonymic rocks. The immediacy of the immeasurably remote past is the Earth's brilliant paradox, or perhaps it is more appropriate to attribute the brilliance to Lyell.

The geological record manifests deep time, making deep time accessible to even the most amateur geologist. Lyell describes how time's spatial dimension, unlike abstract temporality, can be measured, read, and interpreted. He writes:

> By the geometer were measured the regions of space, and the relative distances of the heavenly bodies—by the geologist myriads of ages were reckoned, not by arithmetical computation, but by a train of physical events—a succession of phenomena in the animate and inanimate worlds—signs which convey to our minds more definite ideas than figures can do, of the immensity of time. (1830, 73)

The parallel structure in this passage suggests that as space can be measured so, too, can time be measured. Observation of "a train of physical events" reveals "signs," which in turn suggest "the immensity of time." Time is measurable because change is measurable, but Lyell reminds us that we deal only with signs of time—traces—not with raw time itself. An example from math may prove useful here: mathematicians can work equations involving infinity because they can manipulate the symbol for infinity (∞), not because they are actually able to compute infinity. Similarly, Lyell's assertions about interacting with deep time rely on the symbols of deep time—rocks, bones, and fossils—and on how they reveal change. Furthermore, those symbols are indeed physically manifest; they are material objects that can be excavated and examined by the geologist. Thus, it is through the physical act of excavation that the geologist metaphorically touches time.

The story told by deep time's traces challenges both the biblical creation myth and the notion of narrative. In Scholes's and Kellogg's definition (derived from Aristotle's), narrative depends on "the presence of a story and a story-teller" (4) and "requires . . . a beginning, a middle, and an end" (211). The geological narrative, according to Lyell, has no beginning and no end. Perhaps most alarmingly, it has no author. The geologist's role as reader thus becomes extraordinarily important. The geologist must work hard to decipher the signs embedded in the Earth—the story. Then the geologist becomes the storyteller. He is a translator working from a fragment. He struggles to interpret the words before him and then endeavors to explain to his reader the messages of the Earth's text. Lyell presents himself as an author at the end of Chapter 4 of the first volume of *Principles*, the last chapter to narrate the history of geology: "Meanwhile the charm of first discovery is our own, and as we explore this magnificent field of inquiry, the sentiment of a great historian of our times [Niebuhr] may continually be present to our minds, that 'he who calls what has vanished back again into being, enjoys a bliss like that of creating'" (74).[28] The geologist regenerates the past in the present as he translates fossils and narrates Earth history.

Klaver comments on this authority to create when he describes Lyell "playing God" (45) midway through the first volume of *Principles*. He quotes an extended passage in which Lyell usurps the language used in Genesis; he offers a series of phrases that use the "let . . . be" structure so familiar from the creation story:

> But *let* the configuration of the surface *be* still further varied, and *let* some large district [. . .] such as Mexico [. . .] *be* converted into sea, while lands of equal elevation and extent are transferred to the arctic circle [. . .] *let* the Himalaya mountains, with the whole of Hindostand, sink down [. . .] and then *let* an equal extent of territory and mountains of the same vast height, stretch from North Greenland to the Orkney islands. (qtd. in Klaver 45; Klaver's emphasis)[29]

Klaver then shows that Lyell continues to assume the voice of the creator in a passage that comes near the end of this chapter on temperature changes—"We might first continue these geographical mutations, until we had produced as mild a climate . . ." (qtd. in Klaver 45)—but close examination of this portion of Lyell's text reveals that he speaks in the voice of the maker of the Earth somewhat earlier. Lyell offers a subtly clever syntactic strategy: he begins a new paragraph with a comment about what has come before, "As we have not yet supposed any mutations to have taken place . . ." (120), and then follows with an independent clause that seamlessly transforms the narrative "we" to a creative "we": "we might still increase greatly the intensity of cold, by transferring the land still remaining in the equatorial and contiguous regions, to higher southern latitudes" (120). He continues in this same voice, a scientist's objective first person plural that doubles quite purposefully as the voice of the creator: "The globe may now be equally divided, so that one hemisphere shall be entirely covered with water . . . while the other shall contain less water than land" (121). Here his use of "shall," generally used with the first person rather than with third, further emphasizes the agency of the creator, in this rhetorical instance, himself.

In volume 3 of *Principles*, Lyell extends the agency of creation beyond his own first-person perspective to include the whole geological community. In a section of the second chapter titled, "Origin of the European Tertiary Strata at Successive Periods," Lyell does not describe the development of the strata as the title indicates; rather, he narrates the recent history of geology, recounting how different scientists contributed to the body of knowledge about these strata. Geologists are the subjects of his sentences in this section: "Mr. Webster found these English tertiary deposits to repose. . . . Brocchi traced them . . . ." (18). His language grows more interesting as the chapter progresses. He moves logically from reporting a history of discovery when his title had promised a history of creation to conflating these two: "Mr. Conybeare . . . placed the crag as the uppermost of the British series" (19–20). The verb "placed" allows for the dual action of contributing to a classification system and of literally setting a layer of rock in place.

I would like to turn now to Lyell's discussion of a particular geological site, one that allowed him to conjoin his emphases on uniformity and on the geologist. The constancy of uniformity uplifted the weary spirits of those confronting a rapidly changing world, of those faced with a temporal abyss stretching infinitely back into the past. Sounding the depths of the past yet sublimely present, uniformity soothed existential anxiety. Lyell's reading of one particular relic illustrates the comfort he derives from uniformity: he opens *Principles* with the image of a relic that signifies both uniformitarianism and the creative power of the geologist. His frontispiece is a drawing of the ruins of a third-century Roman temple at Puzzuoli, near Naples. (See figure 2.3.) Borings of marine organisms halfway up the columns indicate that the temple was once beneath the sea. This picturesque ruin provides Lyell an excellent opportunity to conjoin his discussion of cyclicly shifting sea levels with an allusion to a defunct religion, subtly suggesting the fate of the cosmogonists who promote other far less scientific flood stories. Presumably, the temple was built on solid ground, but the water later rose to submerge the structure. The waters have since subsided. Thus, the temple presents the cycle of uniformitarianism in miniature. Lyell states boldly: "This celebrated monument of antiquity affords, in itself alone, unequivocal evidence, that the relative level of land and sea has

Fig. 2.3 The Temple of Serapis, frontispiece to Charles Lyell's *Principles of Geology*, vol. 1 (1830).

changed twice at Puzzuoli, since the Christian era . . ." (1830, 449). In an impressive reversal of the cooption of geological evidence to support the biblical Earth history, Lyell uses a religious relic to advance his principles of geology.[30]

Key among these principles is the notion that pervasive, albeit minute, change rends all features of the Earth inconstant, save one. Water is constant, its movement infinitely uniform. In his extended discussion of the temple, Lyell looks beyond the changing sea level to claim that the level of land may also fluctuate. He argues that the land has fallen and risen again since the construction of the temple, only two thousand years ago, and revises the fundamental essences of rock and water: ". . . it is time that the geologist should in some degree overcome those first and natural impressions which induced the poets of old to select the rock as the emblem of firmness—the sea as the image of inconstancy" (459). Though the agent of much geological change, water is impervious to the passage of time. Lyell explicitly links the sea with the constancy of uniformity when he quotes from Byron's *Childe Harold* (1812): the poet addresses water, "thou / unchangeable" and asserts, "Time writes no wrinkle on thine azure brow; / Such as creation's dawn beheld, thou rollest now" (459). The temple of Serapis, too, seems constant. After thousands of years of shifting land and eroding sea, the remains of the temple still stand, a monument both to the long-dead Roman empire and to the culture of marine life that briefly colonized the Roman relic—a wonderful example of a trace, an artifact from the past marked by the passage of time. Monuments paradoxically work to keep the memorialized alive while simultaneously signifying death. Thus the temple, for all its resilience, represents the death of an empire and its religion. In contrast, the sea exists outside the influence of time, unlike human creations and even the Earth itself. The sea near Naples surely looks and behaves in 1830 much as it did in 30, and Lyell takes solace in this uniformity.

The implications of deep time for humanity are always at the margins of Lyell's texts. The sea is the only unchanging monument, human history offers only a fragmentary example of deep time in miniature, and Darwin's extension of Lyell's logic looms just ahead—the human species itself will probably pass away in the course of time. Lyell ends volume 1 with a chapter on Earthquakes, and his concluding statement puts individuals in their place:

> [The earthquake], so often the source of death and terror to the inhabitants of the globe, which visits, in succession, every zone, and fills the Earth with monuments of ruin and disorder, is, nevertheless, a conservative principle in the highest degree, and, above all others, essential to the stability of the system. (479)

The stability of the system thrills Lyell and overwhelms any component of the system, be it a person or an entire society of people. Lyell offers the rise and fall of the Temple of Serapis as an assurance to the reader: that the culture that built the temple dies along the way does not matter—only uniformity matters.

But uniformity may matter less to the reader than to Lyell. The constancy of process may not assuage the dread of the inconstancy of culture. Beneath the surface of his text, Lyell offers another comfort to assuage the anxiety of deep time. If uniformity itself is not solace enough, Lyell offers the reader the power of the geologist to become deep time's storyteller. Uniformitarianism depends upon understanding the present: indeed, most of Lyell's three-volume *Principles* reads assorted geological phenomena of the present with the recurrent extension of those processes to the past. The geologist-reader of phenomena or traces plays a central role in Lyell's uniformitarian narrative: uniformity may author time, but the geologist narrates the story. In at least an artistic sense, the geologist is the creator.

At the left margin of the frontispiece, a contemporary figure sits on a rock, studying the Temple of Serapis. He and his rock sit in shadow, so we may overlook them, but the observer is a key part of the picture. With his inclusion, the drawing becomes a representation of Lyell's philosophy: the monument has no significance unless someone observes it and translates it; in fact, the observer is part of the narrative. Lyell tells a story about confronting relics and making sense of them. He does not merely enact this form of research, he describes it and invites his readers to participate. The geologist completes the picture of Earth history. He encapsulates deep time in his reading of a trace, packaging time's immensity and disseminating it through popular science. He simultaneously reads and writes the story of time.[31]

"Rid[ing] on the Back of My Iguanodon"

Gideon Mantell focused most of his attention on the observer at Lyell's margins. He both assumed that role himself and wrote five introductions to geology in which he offered his reader meticulous instruction in how to become a geologist.[32] In Mantell's books, the geologist has unequivocal authority. He manipulates fossils and reconstructs them into a cogent narrative. Mantell writes that through geology ". . . we are enabled to link together the phenomena of past and present times" (1848, 26). Mantell's assumption of the uniformitarian view is clear, yet he emphasizes not constant processes but the geologist who observes them. Note that Mantell places the geologist—"we"—in subject position, while past and present are only objects of a prepositional phrase; the syntax reveals Mantell's true subject.[33] Later in the introduction to *The Wonders of Geology*, Mantell explains, ". . . we must remember that time and change are great only with reference to the faculties of the beings which note them" (30). Without the geologist to perceive the past and its connection to the present, time would mean nothing.[34]

Lyell first met Mantell through a volcanic rock he unearthed while doing some casual excavating in a quarry near Lewes. Upon inquiring about the rock, Lyell was referred by the quarrymen to Mantell, whom he immediately visited at his home, introducing himself as a student of William Buckland and thus initiating a conversation that would last many years.[35] As Leonard G. Wilson writes, "the two young men took to each other at once" (93). Wilson goes on to speculate that Lyell began writing *Principles* while visiting Mantell in Lewes in December of 1827 (184). The two corresponded frequently, discussing theories and collaborating on occasion. While Lyell primarily articulated the philosophical principles of geology, Mantell enacted those principles for a popular audience. Lyell even seems to have cast Mantell as his practical counterpart: Wilson writes, "At first [Lyell] had planned a light popular treatment along the lines of Mrs. Marcet's *Conversations in Chemistry* [1806] but he soon found that his own intentions were more serious. . . . He urged Mantell to write a popular book on geology in his place" (186). Eventually, Mantell followed Lyell's suggestion: Dennis Dean asserts that *The Wonders of Geology* (1838), for one, was "the best selling popular geology [book] in Victorian England" (166).[36] Before turning his attention primarily to popularizing not only the principles but also the practice of geology, Mantell was a world-renowned fossilist. He was most recognized for his discovery of the iguanodon, the first dinosaur identified in England, which won him a position in the Geological Society and also the attention of several aristocratic patrons, among them Prince Albert. After selling his personal collection to the British Museum, Mantell turned his attention from fossilizing to writing, in which arena he was also very successful. Both Lyell and Mantell asserted control over the newly perceived expanse of time through the deployment of narrative. Through informed reading of rocks, fossils, and bones, the geologist became first interpreter, then translator, and finally narrator or even author of the Earth.

The frontispiece to *The Medals of Creation* underscores the vital role of the geologist in Mantell's approach to time. (See figure 2.4.[37]) The drawing represents a geological still life consisting of several objects clustered together, among them an ammonite, a fossilized seaweed, a shell, and an ancient sea urchin. These typical "wonders of geology" might have been found in any collector's cabinet. That they are common objects suggests Mantell's effort to reach a broad rather than a specialized audience. He wants people to recognize the objects. Perhaps they already have similar items at home; certainly these fossils are attainable, and the implication is that the practice of geology is attainable, as well. In fact, the practice of geology is illustrated in the drawing: clustered along with the fossils are a geological hammer, a sketch of the geology of Tilgate Forest in Sussex (where Mantell's iguanodon was discovered), a small piece of display board, and, most interesting of all, a pen. These geologists' tools are displayed as objects along with the fossils. The hammer encourages the reader to consider the world of fossils yet to be discovered. Below the lines of strata in the geological sketch we see the name: "G. A. Mantell." The frontispiece, with most of the drawings in the book, was drawn by Mantell's daughter, Ellen Maria Mantell. Thus, the signature within the drawing is not intended to be the artist's mark. Rather, it is a part of the still life, a still life that

Fig. 2.4 Untitled geological still life, frontispiece to Gideon Mantell's *The Medals of Creation* (1844). Courtesy of Special Collections, University of Virginia Library.

features Mantell's signature alongside his signature accomplishment. The geologist's notes and his analyses—even the geologist himself—are objects of study.

The pen, on the left side of the illustration, elaborates on the work to be done. Surely more fossils are to be found, but they must also be analyzed. With the pen, the geologist documents his reading of the past and thus enables the public to read the past through his text. Mantell devotes considerable attention in both texts, but especially in *Medals*, to the methods of the geologist. He instructs his readers in how to find fossils and how to recognize them, but then he goes on to describe ways of displaying the fossils and making them available to others. Among the items he lists as essential for a geological expedition are blank memorandum books (1844, 887). The pointed hammer enables the geologist to extract the fossil from the cliff, but without a pen and paper, how will he make any sense of the fossil? For Mantell, the geologist and his acts of reading and interpretation are as essential as the fossils themselves—the ammonite only becomes an artifact when the geologist pens its description. Mantell's emphasis thus offers a variant on the trace: the trace is a product of eons of accumulated meaning, but it is the geologist who makes that meaning legible. The geologist and his finds depend upon one another, and the story told by Mantell's drawing is the story of reading and writing time.

Mantell was an avid collector and did much to encourage collection among his readers, yet his sense of collecting emphasized possession rather than the objects possessed. Mantell spent his early years as a geologist amassing objects for his own museum. This museum offered a visual narration of time: murals set fossils and bones in their historical context and invited readers to step into prehistory. An anonymous sonnet published in the local gazette in 1836 and reprinted in Mantell's journal described the museum as "a scene sublime / that seems to spurn the bounds of space and time" (134). Baudrillard's assertion that a collection has its own time is apropos here. In an effort to collapse geological time into the space of his museum, Mantell crafted an alternative temporal narrative, one that trumped the reality it metaphorically depicted. In his biography of Mantell, Dennis Dean describes the museum in detail: it was first housed in a special room Mantell had built atop Castle Place, his home in Lewes, for the specific purpose of being a museum; it had six specially designed display cases that held items ranging from minerals to remains of large prehistoric reptiles, like the iguanodon (97). In 1833, Mantell relocated to Brighton, where he reopened his museum in his drawing room.[38] Within three years, the museum operated under the auspices of the Sussex Scientific and Literary Institution and Mantellian museum; in its institutionalized form, the museum was open six days per week and had enough business to require and maintain a curator, George Richardson, who also assisted Mantell with some of his later writing. A popular spot for those on holiday in Brighton, the museum received 150 visitors per week in 1836 (Dean 157), among them most of the celebrated geologists of the day including Charles Lyell and Louis Agassiz and other notable guests, such as the artist John Martin.[39]

In *The Culture of English Geology, 1815–1851*, tellingly subtitled *A Science Revealed through Its Collecting*, Simon J. Knell emphasizes the centrality of the museum for nineteenth-century geology. He claims, "In its use of fossils, geology was to become as much a science of the museum as it was of the field" (95). Though Mantell certainly spent time in the field, it was his collection—his museum—that formed the foundation of his reputation. I should note, however, that praise for Mantell's museum was always implicitly or explicitly praise for his skill as a collector and interpreter of fossil remains, for nearly all of the items on display had been collected by Mantell himself. In his 1829 review of the museum, Robert Bakewell (1768–1843) clearly links Mantell the collector to Mantell the curator:

> Many of the specimens in this collection are unrivalled and unique; indeed, we are entirely indebted to the scientific investigations of Mr. Mantell, for the first knowledge of their existence. . . . When Mr. Mantell first commenced his researches in the vicinity of Lewes, no fossil organic remains had been collected there, nor had the quarry men noticed them in the beds they were daily working, but in the course of a few years, Mr. Mantell, succeeded in obtaining the finest collection of chalk fossils in the kingdom. (9)

The museum is thus not only a space to display the remains of geological time but perhaps more importantly the evidence of Mantell's skill as collector and interpreter of these remains.

The museum, or any even slightly organized collection, presents traces of the past in a present-day context that offers an interpretation, whether it is implicit or explicit, of those traces.[40] A geologist may interpret a fossil or a geological feature of the landscape in the field; for instance, he may observe the placement of strata and draw conclusions about the order and duration of epochs. But the geologist removes the object from its context and crafts an artificial context that is itself an interpretation when he puts the object on display, and any person, however ignorant of the principles of geology, can make sense of a fossil when it is located in a sense-making environment, like a museum. Fragments are reassembled and joined with fabricated faux fragments to form a whole skeleton, thus presenting to the interested viewer a clear image of a dinosaur or other prehistoric beast; fossil fishes are laid out in drawers, one drawer displaying the most primitive fishes, the next drawer displaying more advanced fishes, and so on, so that the visitor can peer into the past and observe the evolution of fishes through the simple, even domestic, act of pulling open drawers. Discussing the Scarborough Museum, Knell describes the influence of the eminent stratigrapher William Smith on the layout of objects within the museum. He quotes an anonymous review: "The fossils, which are very numerous are arranged on sloping shelves, in the order of their strata, showing at one view, the whole series of the kingdom" (69). This ability to see the "whole series of the kingdom" at one view was empowering to a public who may have otherwise felt overwhelmed by the geological evidence. Barbara J. Black describes this very phenomenon in *On Exhibit: Victorians and Their Museums*: "the self initially collects out of curiosity for an other . . . yet possession ultimately empowers the self because the very act of collecting demystifies the unknown" (22).

Mantell reluctantly sold his collection to the British Museum, after several years of geological success but financial loss. Yet, just as a collection on display offers a visitor a coeval encounter with the past, so might a narrative. Mantell turned to instructing others in the acquisition of geological relics and to rehearsing his own successes as a fossilist in his five popular books. Excavation, collection, and assembly all precede display, and Mantell attends to these stages of the process in several of his introductions to geology. Knell observes that the material elements of geology were widely accessible to all sorts of people and, thus, the act of collection was one anyone could perform, regardless of financial means or scientific background. Fossils "linked labourer to aristocrat, dilettant to savant, gift-giver to local elite" (xi). Mantell worked hard to make both the narrative and the experience of geological time accessible to his readers. A planned expedition in the field was surely the best approach to the past, but in the complicated world of contemporary England, other approaches made geology even more pervasive. The rapidly expanding rail network in England accidentally contributed to the popularity, even the possibility, of geology.[41] To build a railroad bed, the workers sliced into the Earth, revealing layers of rock and sediment, as Mantell describes: "In the works now in progress for a branch railway through Trowbridge, the Oxford clay underlying beds of Coral-rag, has been largely exposed, and has yielded innumerable specimens of the usual fossils, and some species of ammonites and other shells not previously observed" (1848, 502). Mantell's language here is the language of progress and productivity; he conflates the growth toward the future with a growth of the past into the present. His use of the verb "yielded" emphasizes a sense of growth and progress. Like interest on an investment, fossils and the past they signify are the yield on the investment in the future. Ironically, progress leads back in time—progress and decay paradoxically coexist, as Lyell demonstrated.

Here again, Mantell emphasizes the process that revealed the fossils rather than the fossils themselves. He extols the construction of the railroad beds as an invaluable geological tool, and he goes on to offer train travel as a sort of text for a student of geology. In *The Wonders of Geology*, he uses the route of the Great Western Railway to describe the array of strata within a small geographical area. He repeats the same passage almost verbatim in *The Medals of Creation*, but in the later text he aims to provide the reader with a geological journey. Mantell includes the description of the railway among other descriptions of excursions the amateur geologist might enjoy:

> That splendid railway, the Great Western, by which the geologist may be transported in five or six hours, from the Tertiary strata of the metropolis, to the magnificent cliffs of Mountain limestone at Clifton, exposes in its course several fine sections, and passes within a moderate distance of some interesting localities of organic remains.
>
> This railroad traverses the *Tertiary* strata by Ealing, Hanwell, and Slough, entering the *Chalk* near Maidenhead, and pursuing rather a circuitous route to Wallingford, beyond which station it passes over the *Oolite*, and displays some bold sections of the limestones and clays of that formation. Near Bath it emerges on the *Lias*, and crossing a narrow belt of the *New Red*, passes on to the *Carboniferous* strata of the Bristol coal measures. (1844, 922)

This short course in geological epochs reveals the locations of assorted strata and makes sense of the depth of time by setting it in the context of familiar contemporary England. The expanse of time becomes less overwhelming when the distance between the Tertiary and Chalk eras can be represented by the distance between Ealing and Maidenhead. Mantell employs a spatial synecdoche, using finite geographic space to represent the infinite abstraction of time.

The railroad offers a useful teaching tool for the geologist, but Mantell actually proffers the Great Western Railway as a route to the distant past. In the second paragraph, Mantell accompanies only the first and last epochs with the word "strata." After leaving the "*Tertiary* strata," one actually arrives at the "*Chalk*" and then moves on to the "*Oolite*." Leaving off the word "strata" creates the impression that the train travels back in time. Mantell's reference in the first paragraph to the length of the journey—"five or six hours"—foregrounds temporality. In the five hours it takes to travel from London to Bristol, one could travel from the Tertiary period to the Carboniferous. The ease of a train journey is equivalent to the ease of the geologist's journey to the past. By moving physically through the space of England, a person can journey from one epoch to the next, stopping at distinct strata as one would stop at so many train stations. The train journey Mantell describes offers a striking example of anachronistic space: the passenger-reader watches the Earth-text move past the window, and he thus understands time's story. The train becomes the vehicle for the geology Mantell authors.

Train travel allows any passive amateur geologist to visit epochs past, but Mantell devotes most of his writing to more active acquisitions of traces. He concludes *The Medals of Creation* with a section devoted to the logistics of encountering time, "Notes of Excursions." Mantell intends these "notes," as the lengthy subtitle explains, to illustrate "the mode of investigating geological phenomena," and, while they surely accomplish such illustration, they really comprise a travel guide. Mantell recommends particular inns, where the geologist-traveler can find knowledgeable proprietors, some of whom even offer local fossils for sale. He also suggests itineraries that include nongeological sites. For example, Mantell's description of Crich Hill in Derbyshire concludes with a recommendation to visit a nearby historic ruin that has no geological significance whatsoever:

> A long summer's day is not too long to visit [Crich Hill], and examine all its interesting details. A good pedestrian should proceed with his hammer, and haversack, for every step of the road is replete with interest; and as numerous specimens will be obtained, bags, paper, and boxes should be taken . . . and if time permit [*sic*], the interesting ruins of South Wingfield Manor-House (once the prison of Mary of Scotland) about two miles from Crich, may also be visited. (1844, 952)

Mantell conjoins exploration of a past millions of years distant with a tour of a sixteenth-century home. He makes a similar juxtaposition when he describes the geology near Chepstow and the ruin of Tintern Abbey (1844, 930), a conflation of geology and wrecked religion rather like Lyell's Temple of Serapis. His suggestion of combined geological and historical excursions reflects a notion of coeval time: just as geological epochs are ranged alongside one another in the form of strata, so, too, do geological and human pasts lie side by side. Mantell's audience consists of geological tourists, who will journey from London to Derbyshire, and, as easily, from 1843 to 1580 to two million BCE.

While Mantell takes pains to prepare his reader for the practice of geology, his primary interest seems to be a celebration of his own contributions to the young science. His description of his supposed discovery of the first iguanodon illustrates the extent of his authority in geology and over time. Describing Mantell's reputation in "A Visit to the Mantellian Museum at Lewes," Robert Bakewell describes Cuvier's endorsement of Mantell's find: "This illustrious anatomist pronounced the Iguanodon, discovered by Mr. Mantell, to be a reptile more extraordinary than all those which have been hitherto known" (10). Mantell writes at great length in his first three books about the iguanodon, celebrating his own prowess. He describes his reconstruction of the disjointed skeleton and pays particular attention to his decision to name the monster for the similarity of its teeth to those of the modern iguana. His self-conscious assertion of his own role as the author of the iguanodon gives him considerable authority in professional geological circles but also in his own narrative of Earth history. He authors the past when he tells the story of the iguanodon; he even makes its literal descendent into the creature's namesake. This temporal confusion reflects the power Mantell wields over this trace from the remote past. In the face of something so old, one might feel very small, but Mantell makes himself bigger than the iguanodon. By conquering the fragmented traces of the iguanodon, he conquers and controls time. He writes in his journal of September 1829, "I shall ride on the back of my Iguanodon into the temple of Immortality" (72).

Mantell could not possibly have anticipated the way in which his iguanodon would itself achieve lasting renown. In 1853, one year after Mantell's death, the iguanodon became the site of an extraordinary inaugural event for the Crystal Palace at Sydenham.[42] The grounds of the palace contained a dinosaur park, an arrangement of full-scale plaster cast prehistoric reptiles. Constructed by Waterhouse Hawkins (1807–1899) from anatomical drawings by Richard Owen, the models were much praised for their detail and accuracy and offered as close an encounter with the past as many Victorians could achieve. The closeness of such encounters was emphasized by a notorious dinner held on New Year's Eve 1853, inside the model of the iguanodon. Figure 2.5 shows a drawing from the *Illustrated London News*. The dining gentlemen, among them renowned paleontologist Richard Owen, are at the center of the illustration, but the detail of the iguanodon's face, immediately to the left of center, reminds the viewer exactly where these celebrated men sit. This "socially loaded stomach" (*Illustrated London News*) conflates the present of 1853 with the past so precisely represented by the model. In fact, the iguanodon is imagined to have come to life in the present and quite enjoyed himself. The *Illustrated London News* concludes its report on the event with the following humorous but illustrative speculation: "After several appropriate toasts, this agreeable party of philosophers returned to London by rail, evidently well pleased with the modern hospitality of the Iguanodon, whose ancient sides there is no reason to suppose had ever before been shaken with philosophic mirth." The fierce primordial past could not be more tamed than it is on the occasion of this comical dinner. In her discussion of this famous event, Barbara Black references a *Punch* article that highlights the ability of the present to tame the primeval past: "An article in *Punch* praised 'the company on the era in which they live; for if it had been an early geological period, they might have occupied the Iguanodon's inside without having any dinner there.' Victorian science had, in effect, not only rescued but revised history, replacing the ravages of nature and animal with the triumphs of culture and humankind" (22). She goes on to explain the event in the context of montage, the re-creation of an object—in this case, a dinosaur—from fragments. She explains, "Owen's reconstruction of the Iguanodon was culturally determined and socially significant—a whole beast, an entire species based on fragmentary knowledge" (23). The Crystal Palace dinosaurs, though entirely artificial, thus follow the tradition of reconstruction that Mantell practiced and endorsed. With the iguanodon dinner, the Victorians demonstrate that they can reconstruct the past and domesticate it, asserting their own authority over the past and assuaging the anxieties kindled by the prehistoric monsters who give such a brutish face to deep time.

A poem by Horatio Smith (1779–1849), "A New Nightmare," is set in the Mantellian museum just before the contents were shipped to the British

Fig. 2.5 "Dinner in the Iguanodon Model, at the Crystal Palace, Sydenham," *Illustrated London News* (1854).

Museum and dramatizes the horror implicit in geological fragments, a horror that is contained when those fragments are assembled into dining rooms or at least whole creatures. Smith begins, "Doctor Mantell's Museum was all disarranged" (1); the premise is that such disarrangement is a nightmarish setting. The removal of geological specimens from their display cases is clearly threatening:

> The huge fossil bones on the carpet were laid,
>   To be packed for conveyance to London;
> The Curator was gone, there was no other guard,
> All the drawers were unlocked, all the doors were unbarred,
>   All the cases were open and undone.
>
> (2–6)

The speaker falls asleep "midst these relics" and in his dream is transported to "the great geological eras" where he encounters, "Realities far more terrific and grim / Than the wildest of fabled chimeras" (11–12). What causes such terror is the reanimation of the "monsters" who assume their complete forms and descend upon the frightened speaker, accusing him of having removed their bones from their resting places and crudely displaying them to "featherless biped beholders" (36). The speaker desperately defends himself: "I'm not Dr. Mantell—I swear to the fact" (38). The prehistoric animals do not believe his oath, however, and they attack him: he is "baited and gored / by the whole of the antediluvian horde" and then his bones are smashed by the tail of the iguanodon. When the speaker wakes from this nightmare, he is relieved to find himself back in the museum. The poem is obviously humorous, but it articulates one of the core principles advocated by Mantell: reconstruction is a means to contain and control the threatening past, threatening in its fragments, which signify not only the expanse of the past but also the likelihood of the future.

The irony of Mantell's creation of the iguanodon, however, is that it seems he might not have discovered this particular fossil at all. Secondary sources credit Mantell with the discovery, but Mantell's journal suggests that he purchased the skeleton. Robert Bakewell asserts that the first iguanodon teeth were in fact discovered by Mantell's wife, Mary Ann (13). It is possible that there were two early iguanodons, or that one was complete and the other constructed out of fragments. In any event, one early iguanodon languished in the possession of a local man until 1834 when Mantell tried to buy it. His bid was rejected, but several of his friends took pity on him and purchased the specimen on his behalf. Mantell did not discover this iguanodon, nor did he even successfully broker its purchase. The passivity of his acquisition combined with the authority he assumes over the iguanodon indicate that possessing the object and telling its story are far more important than discovering it.

For Mantell and other Victorian geologists the field became less a laboratory than a marketplace. Four years before acquiring the complete iguanodon skeleton, Mantell attempted to purchase a jawbone to contribute to the construction of a skeleton built from scattered fragments. The jaw was uncovered in a stone quarry that doubled as a fossil quarry for Mantell. He describes his frustration with the fossil market in his journal:

> Drove to Cuckfield, and endeavoured to obtain some fossils from the quarrymen who have been employed by me so many years; but the ungrateful scoundrels refused to let me have one, having found a customer on the spot; a gentleman who has just become amateur and collector: here there is an end to all my hopes of discovering the jaw of the Iguanodon. It is after all a very cruel affair: when there are thousands of quarries in England which would be equally productive and interesting if persons would but take the trouble to explore them, I must be robbed of the fruits of my industry and trouble! But so I have always found it. (89)[43]

The quarrymen were Mantell's fossil agents; they uncovered relics on his behalf and sold him their finds. This geological business enhanced Mantell's sense of ownership. When his agents failed him, he became indignant. His power over geological time depended on his ability to compete for fossils in an open market. Paula Findlen describes this commercial aspect of natural history:

> While naturalists saw themselves as members of the republic of letters, they were dependent upon men who knew nature by profession to complete the unwritten chapters in their new histories of nature. Initially, the 'book of nature' was far removed from the humanists' *studio*. The men who possessed it were neither physicians nor philologists but tradespeople who understood nature to be a commodity upon which their economic survival depended. (Findlen's emphasis; 171)

Findlen goes on to comment on the evolution of knowledge away from the pristine sphere of the university into the more real-world sphere populated by a variety of experts: "Classical wisdom was increasingly tempered by the untutored descriptions of apothecaries, gardeners, fishermen, and hunters whose occupations gave them privileged access to the secrets of nature" (175). Mantell does not give his quarrymen any intellectual authority, but he is frustrated with them because they have the power to facilitate or impede his own authority.

We see the commercialization of geology and deep time in the active fossil dealing that occurred in seaside towns and in London. *The Medals of Creation* concludes with a list of London's fossil dealers, dealers who also advertised regularly in the popular magazine *The Geologist*.[44] These merchants procured fossils—along with minerals, rocks, and bones, in addition to books, maps, charts, and plaster reproductions of everything from prehistoric crustaceans to life-sized dinosaurs—and sold them to collectors who could not or would not go out into the field to acquire relics on their own.[45] One fossil merchant in London, James Tenant, advertised collections intended to accompany popular texts by Mantell and others. For just over one pound, a collector could buy "fourteen models, carefully coloured from the originals, of Teeth and Bones of the Iguanodon . . . discovered by Dr. Mantell" (*Geologist*, n.p.). Readers could purchase collections to complement texts and texts to complement collections, such as Tenant's guide to his own inventory. Tenant also sold cabinets full of actual geological remains; he advertised a twelve-drawer cabinet with four hundred specimens for twenty-one pounds. Purchasing such a cabinet would instantly establish the buyer as a successful collector, one who may never have gone on a field excursion but who nonetheless recognized the value of possessing a fossil collection. Barbara Black asks the provocative question: "What does one do with all the things of the world?" (15). The answer for Tenant and his customers was simply purchase and display them, whether real or artificial, and thus contain them.

Outside London, the most renowned dealer of the early nineteenth century was Mary Anning (1799–1847). This peculiar woman from Lyme Regis figures prominently in the journals of both Mantell and Lyell and in the writings both personal and professional of just about every geologist of the early nineteenth century.[46] Anning took over her father's small fossil business when he died in 1810, and she eventually became so skilled at discovering valuable fossils that she supported her mother and siblings through her fossil-dealing business. She excelled at fossil discovery and identification and developed a reputation that even in her lifetime bordered on the legendary. Her renown was described in "Mary Anning, The Fossil Finder," an 1865 article published in *All the Year Round*:

> But Mary Anning was something more than a mere village celebrity, interesting to those who like to study character, and are fond of seeing good stubborn English perseverance make way even when there is nothing in its favour. . . . Professor Owen thought so highly of her usefulness, that he moved the authorities of the British Museum to grant her a pension of forty pounds a year, which she enjoyed for some little time before her early death. (61)

Several historians of science have observed that with this pension Anning was the first woman to receive government support for her scientific work. However, in the context of mid-Victorian geology, measuring Anning's value in fiscal terms was charitable but hardly complimentary. She was not so much a scientist as a businesswoman, one who was respected for the quality of her wares and the depth of her knowledge, but a businesswoman nonetheless and little more.[47]

As Knell argues, fossil collectors were divided into two camps: the gentlemen geologists and the tradesmen fossilists. This division has a great deal to do with the dual origin of the young science of geology: growing out of studies to improve mining and simultaneously developing out of the philosophical societies, geology had ties with philosophy and labor from the beginning. Those who earned money from the fossils they discovered were in a different class altogether from those who made occasional forays and wrote the defining literature of the new science. What's more, as Mantell's case illustrates, the gentlemen geologists often based their conclusions on fossils purchased from geological tradesmen. Though she excavated the traces of deep time, Anning had no authority to write time's story. The very few literary traces that Anning did leave behind reveal little about her and do little to claim any authority in the world of geology. Her letters consist almost entirely of business transactions: she presents her merchandise, discusses its value, and proposes a price. For instance, she wrote to Mrs. Murchison in 1829: "I have got a very good head of an Ichthyo vulgaris about two feete in length which I value at five £—and this day I have found a beautiful Ammonites Obtusus about a foote across which I value at one pound, pray do you think it will do for Mr. Lyle [sic]" (qtd. in Lang 170). Anning's only publication was a short note, "Note on the Supposed Frontal Spine in the Genus Hybodus," published in the *Magazine of Natural History* in 1839. It was evidently written to clarify a confusion and is most interesting for the way in which Anning references important geologists, De la Beche and Agassiz, and then classes herself with them with her use of first-person plural. If there were more publications like this one (or even any more of any kind at all), I might argue that Anning did assert a literary authority over the traces of time as did Lyell and Mantell, but this one brief note does not offer the basis for such a claim.[48] For Tenant and Anning, as well as Mantell's quarrymen, the past became the raw material of present industry, and the trade in traces was a profitable venture, in terms of monetary gain if not scientific authority.

The past also became the raw material of present buildings and monuments, a marvelous illustration of traces that interested Mantell in *The Wonders of Geology*. Remains of the past have a continued life in the present not only as singular items to display in museums or drawing rooms, but also as practical and aesthetically pleasing building material. Mantell writes that "the beds of fossil coral are now the sites of towns and cities, whose inhabitants construct their abodes of the limestones, and ornament their temples and palaces with the marbles, formed of the petrified skeletons of the zoophytes, which lived and died in oceans that have long since passed away" (1848, 658–69). Ricoeur could not have asked for a better example of the trace. Just as scientists reconstruct fragments into whole models of past creatures, people build the structures of modern life out of these same fragments. Charles Kingsley (1819–1875) pursues this theme in great depth in *Town Geology* (1872). He describes the pebbles in the street in geological terms, explaining that they are 37 percent igneous rocks, 43 percent Silurian grits, and so on (65). He devotes one chapter to lime, which forms the mortar between bricks in the walls of the reader's home, and another chapter to slate, which comprises the roof above the reader's head. In addition to buildings constructed from ancient materials, monuments also rest on fossil foundations. Mantell describes the many fossilized sea creatures that comprise the marble of contemporary monuments. Thus, remains of distant epochs become the raw material for the remains-to-be of the present epoch. A fascinating example is the gravestone of Lyell, who was buried in Westminster Abbey in 1875. Lyell's bones lie beneath a marker of fossil marble from Derbyshire, selected as an appropriate tribute. The fossil creatures in the marble pay homage to the fossilizing geologist buried beneath. The modern world is constructed with the vestiges of deep time, vestiges that have a new life as they are employed in homes, places of worship, and political establishments.

Beyond buildings, monuments, and even individuals, the story of time becomes the story of the British Empire. Iguanodons and mammoths are described as citizens of the empire in a caption beneath an illustration of the reptile park at the Crystal Palace—"Struggles of life among the ancient Britons in Ante-Diluvian times" (*Geologist*, n.p.). Mantell even refers to London as the capital of the Tertiary period: "The bed of an ancient sea, containing myriads of the remains of fishes, crustacea, and shells, now forms the site of the capital of Great Britain" (1848, 286). Mantell's diction calls attention to the notion that the past is actually part of Great Britain: along with India and East Africa, the English have colonized the Tertiary period. Mantell furthers this assertion in his 1850 picture book, *A Pictorial Atlas of Fossil Remains*. This atlas enables the reader to navigate temporal depths, moving not only latterly about the globe but also plunging into the depths of the Earth, and the depths of time. Just as the Victorian geographic atlas suggests that distant places had been conquered, a geologic atlas suggests understanding and conquest of the inhabitants of distant times and, by extension, conquest even of those past times. The British Empire thus operates across an infinitely vast temporal dimension in addition to the spatial dimension of the globe. Such presumption surely reflects the scope of imperial ideology in this period, but it also reflects assurance in the power of the geologist. In his introduction to *Rule of Darkness*, Patrick Brantlinger explains that imperial ideology was a significant element of the national psyche at mid-century. He writes of Trollope, "he believes that, wherever the British flag flies, he and his compatriots have a responsibility to import the light of civilization (identified as especially English), thus illuminating the supposedly dark places of the world" (8). Mantell and his fellow geologists surely worked to illuminate dark places—dark due to depth of space and time, rather than matters of race or civilization—and if Mantell could have raised the Union Jack in the Tertiary, he would surely have done so. The British government metaphorically claimed geological time for England with the creation of the Geological Survey in 1835. Like military imperial ventures,[49] the work of the Geological Survey was undertaken by teams of geologists working in coordination to fully master a particular region of the country, to conquer it through comprehensive mapping of strata and analysis of fossil inhabitants of each strata.[50] If a single geologist can possess the past in his cabinet at home, then imagine what a nation of geologists might possess.

Mantell speculates about possessing not just the distant past but also the future in both *The Wonders of Geology* and *The Medals of Creation*. In the earlier text, he mentions human skeletons that had been uncovered in limestone. From this point he makes the logical leap to consider the preservation of the human race in the future:

> The occurrence of human skeletons in modern limestone . . . incontestably prove[s] that enduring memorials of the present state of animated nature will be transmitted to future ages. When the beds of the existing seas shall be elevated above the waters, and covered with woods and forests—when the deltas of our rivers shall be converted into fertile tracts, and become the sites of towns and cities—we cannot doubt that in the materials extracted for their edifices, the then existing races of mankind will discover indelible records of the physical history of our times, long after all traces of those stupendous works, upon which we vainly attempt to confer immortality, shall have disappeared. (1848, 124)

In the present landscape, Mantell sees the future: seas of today are the forests of tomorrow and today's rivers are tomorrow's towns. In predicting the geological future, Mantell converts present life into traces: just as the stone-carver of the Victorian period crafts monuments out of fossil marble, so will the stone-carver of the future, only his fossils will not be crustaceans but will be the fossilized remains of men, women, and children of Mantell's own time. It is striking that Mantell does not simply imagine these fossilized remains; he goes the extra imaginative distance of imagining those remains as fragments within building materials. As traces, humans will be constructive material to the future. Writing humans as potential fossils offers an odd sort of immortality. Deep time does not overwhelm human significance; rather, humans have an infinite existence as traces because of deep time.

Historians of archaeology have observed that archaeology and geology were one and the same at these sites of human remains. Hugh Torrens points out that the gulf between geology and archaeology is a modern invention, one which we impose backward to the Victorians. For them, the two fields were quite seamless.[51] In *A History of Archaeological Thought*, Bruce Trigger explains, "This new view of geological history also left the question of the antiquity of humanity as one that required an empirical answer. The favourable reception given to Lyell's geology reflected the increasing openness of British scholars and the public to evolutionary ideas" (93). Trigger discusses how archaeology took its first methodologies from geology in its reliance on dating strata by the artifacts discovered there. The other side of the coin, however, is that the discovery of human remains as geological material reinforced the anxieties about species extinction and the ultimate insignificance of human life. I discuss how archaeology sought to rescue the individual, the species, and even the culture in chapter 4. Indeed, Mantell seems buoyed rather than depressed by the presence of human remains amongst geological relics.

Mantell describes human creations as traces-to-be in more detail in *The Medals of Creation*:

[The strata's] most striking features would be the remains of Man, and the productions of human art—the domes of his temples, the columns of his palaces, the arches of his stupendous bridges of iron and stone, the ruins of his towns and cities, and the durable remains of his Earthly tenement imbedded in the rocks and strata—these would be the "Medals of Creation" of the *Human Epoch*, and transmit to the remotest periods of time, a faithful record of the present condition of the surface of the Earth, and of its inhabitants. (876)

Note that Mantell's emphasis here is on human creations, not on humans themselves. His final parenthetical phrase, "and of its inhabitants," makes the inclusion of human remains implicit in his speculation about the future, but he does not threaten his reader with this alarming eventuality. Mantell did not live to read the *The Origin of Species*, though Darwin was certainly not the first to extend geology's implications into the human realm, and like many of his geologist contemporaries, Mantell surely recognized what species extinction meant for the human species, but he did not make those conclusions explicit in his published works. Indeed, in this passage, Mantell turns his present into the remains to be discovered by the geologists of the future, thus assuming that the future will include human scientists who will practice geology much as it is practiced in the mid-nineteenth century. He imagines that the art, engineering, and architecture of London will be nothing more than a collection of fragments to be unearthed, interpreted, reconstructed, possessed, and displayed by the imperial dealers of the future. London itself is thus cast as a trace. These passages represent a fascinating move on Mantell's part: he discards his practical, scientific method and adopts a speculative, overtly narrative tone. He portrays humans and their cultures as traces awaiting discovery by readers from the future. These readers will know us by our domes and palaces, and we will gaze at them through our fossilized eyes; and past, present, and future will merge into a coeval epic in which dinosaurs and geologists live happily together.

Mantell concludes *The Medals of Creation* with a futuristic short story written by Thomas Hood (1799–1845) in which men from the future participate in a geological expedition. In this story, set in the year 2000, "anticipatory for the 100th edition of the *Medals of Creation*" (1844, 982), a group of men head out to the Tilgate Forest to uncover some bones. Introducing the premise of the story, Hood philosophizes about the accomplishments of geology. What were once rare or even mythological creatures have been made into common-places. Their bones, he writes, are "Mantell-pieces" now (982). The pun on "mantelpiece" signifies how deep time has been domesticated and emphasizes Mantell's role in that process. The bones of an iguanodon have been imported from the Tertiary period to live comfortably in a Victorian, or twenty-first-century, drawing room. Through geology, time has been contained and conquered. Hood goes on to describe such a domestication.

The geologists readily uncover small relics, a "fossil red-herring"—double meaning intended, I think—for example, but these items are mere "geological bagatelles" (983). Finally, they find a large animal, and the rest of the story recounts their gradual discovery of the whole beast and their speculations about its identity. I offer a short sample:

> Bravo! There's his cranium—Is that brain, I wonder, or mud?—no, 'tis conglomerate. Now for the cervical vertebrae. Stop—somebody hold his jaw. That's your sort! There's his scapula. Now then, dig boys, dig, dig into his ribs. . . . There's his tarsus! Work away, my good fellows—never give up; we shall all go down to posterity. It's the first—the first—the first nobody knows what—that's been discovered in the world. (984)

In the next paragraph, the narrator makes the intriguing claim, "we're all Columbuses" (984). He refers to the process of discovery, certainly, but he also refers to colonial expansion. The geologists discover a new creature from the past and promptly colonize that past. They encroach from their present vantage point and with their present-day tools and present-day naming on the very distant past. They narrate the past and make it their own.

The past in this case proves to be even more distant than the past of the Tertiary period: the creature is from the distant, ephemeral past of mythology. The narrator exclaims, "by Saint George, it's a flying dragon" (984). The dragon, with the exclamation to St. George, is a symbol of England: thus, the diggers uncover a piece of their own identity. This discovery legitimizes what had been thought a fairy-tale creature; thus, we must revise our understanding of the real past. Geology enacted a similar narrative with the early discoveries of dinosaurs. That giant reptiles had once walked the Earth was a fantasy at most, and it took a collection of paleontological discoveries to turn the fantasy into reality. So, the finding of a dragon is not so absurd.

What is delightfully absurd is the ensuing parody of Mantell. A "stony-hearted Professor of fossil osteology" examines the dragon's teeth and proclaims that it is a Mylodon, named for its molar teeth: "That creature ate neither sheep, nor oxen, nor children, nor tender virgins, nor hoary pilgrims . . ." (985). This Mantell-esque professor never denies that the creature is a dragon; he simply renames it in a way appropriately reflecting its physical characteristics. Thus, the dragon, one of the best-known figures of mythology, is rewritten. It is reclaimed into the present (actually the future), and all that we believe about dragons is erased by the reality of the creature's teeth. This final scene of the story is comic but also strikes a serious note. In reading the dragon-fossil, the professor gleans insight into the past. His act of reading becomes also one of writing as he reconstructs the dragon, and thus reconstructs the past. In this way, the present generates the past. The story is conveniently set in our present of 2000, but to Mantell and his readers that date represented the future. Thus, past, present, and future come together in a moment of coexistence and mutual authorship.

Nineteenth-century geology introduced deep time and the accompanying principle of uniformitarianism. Uniform processes only affect noticeable change when they are constant over a vast expanse of time. Thus, uniformitarianism depends on the existence and acceptance of deep time. Yet, deep time only becomes legible when uniformitarian geology reads the surface of the Earth and interprets the traces of change. Thus, deep time depends on the existence and acceptance of uniformity. The contingent relationship between the two left many early observers confused. On one hand, deep time is awesome and alarming, making the individual feel utterly insignificant. Tennyson famously mused on the tragic irrelevance of humanity in a world governed by infinite time in *In Memoriam*: he personifies an indifferent nature in lyric 56,

"So careful of the type?" but no.
From scarped cliff and quarried stone
She cries, "A thousand types are gone;
I care for nothing, all shall go."

Yet, on the other hand, the geologist exulted from cliff and quarry over his authority to read the past and report its story. Rivaling the greatest Author of all, uniformitarian geologists stood outside time, observed its scope and its work, extended their observations through imagination, and crafted a narrative of Earth history that ultimately replaced the biblical one. They uncovered remains of a prehistoric, pre-human past, and they made the stories told in the Bible seem small. At the same time, they assured the significance of humans, primarily geologists, as narrators of Earth history.

The geologist's authority threatened many but also comforted those, like Tennyson, who saw the individual wither as time grew deeper and deeper. In his introduction to Robert Chambers's areligious *Vestiges of the Natural History of Creation*, George B. Cheever writes, "This book has been called a scientific romance. It is the most ingenious and elaborate attempt we have ever seen to turn Nature into Fiction, and to exclude God utterly by law, from his own world" (Chambers 1845, vii). Cheever recoils at the rewriting of Earth history, and he disparages geology, which produces, in his view, "romance" and "fiction." Lyell and Mantell believed that their texts represented the facts of the prehistoric world, but they would have readily agreed with Cheever's association of geology with narrative. The observer on the margins, the pen equated with the ammonite—these illustrations reflect the most fundamental of the principles of geology: without the geologist to dig up the fossil, read and interpret its story, and recount its history, the depth of time would go unnoticed, changes so gradual would never be recognized, and Earth history would remain nothing more than a romance. Moved by the sublime landscape in which everything signifies a distant past, geologists charge ahead with a pen and pick axe and write time.

3

Tennyson's
Fairy Tale of Science

> . . . —and that hour perhaps
> Is not so far distant when momentary man
> Shall seem no more a something to himself . . .
>
> —Tennyson, "Lucretius" (251–53)

G eologists were not alone in their efforts to rescue the individual from insignificance. The literature that developed to explain and disseminate the new science of geology merged with the larger discourse that attempted to locate humans vis-à-vis the natural world and that struggled with the inevitable problem of recognizing that humans are not central to the Earth's story. In *Darwin's Plots*, Gillian Beer writes, "New organizations of knowledge are particularly vexatious when they shift man from the centre of meaning or set him in a universe not designed to serve his needs" (14–15). By placing the individual in grammatical subject position and by reading geological traces as texts that require human readers to make sense of them, geologists fought to keep humans at the center of the universe, to make man seem "a something to himself."[1] In this project, they were joined by writers whose subject was only obliquely geology but who urgently attended to the problem of humanity's diminishing centrality. In this chapter, I examine Alfred Tennyson's poem, *The Princess*. First published in 1847, this comic-epic-romance appeared after the broad dissemination of Lyell's *Principles*, but before the public felt wholly comfortable with the theory of uniformity and the necessary expansion of time.

Tennyson tackles many diverse subjects in *The Princess*, and my analysis focuses on an often neglected subject of the poem—the value of the individual life in the face of geological evidence. Tennyson's princess struggles to make a mark on time—to leave a trace—and Tennyson suggests that the struggle for reform must be reconsidered in light of geology. Much has been written on Tennyson's attention to geology and other sciences in *In Memoriam*, but readers of *The Princess* generally focus on other matters, and not without reason. The poem is epic in its scope, and to claim it is a poem about geology would be reductive and outright false. Nonetheless, the subject of geology is important to the poem, and I will show how Tennyson's use of geology inflects the poem's comic resolution.

"Nourishing a Youth Sublime"

Like the geologists, Tennyson was awed by the sublime landscape and the natural history it undeniably revealed. He conveys his mixed feelings about geology in "Locksley Hall" (1842),[2] in which the melancholy speaker recalls a youth "nourished" by geology and the wonders it reveals: "Here about the beach I wander'd, nourishing a youth sublime / With the fairy tales of science, and the long result of time" (11–12). The "fairy tales of science" may describe geological texts or theories or even fossils; the specifics are less important than the magic and delight associated with science through Tennyson's use of the term "fairy tales." "The long result of time" conveys at once the depth of time and also uniformitarianism with the emphasis on "result." The conjunction of the two phrases conjures again the notion of the trace—the items on the beach tell stories of science and indicate the passage of time—and the trace is empowering, or nourishing, for the youth. In contrast, the reflection of the older speaker in the present is bleak: "Knowledge comes, but wisdom lingers, and I linger on the shore, / And the individual withers, and the world is more and more" (141–42). The setting is still the beach, but the mood is entirely different. As a youth, the speaker was buoyed by the wonders of science, but as a mature individual, he feels himself wither away to virtually nothing before the knowledge that the shore and its fossils foist upon his generation. We see in "Locksley Hall" Tennyson's inclinations to marvel at geology and recoil from it in turn. He expresses this vexed attitude in his two great poems of mid-century, *The Princess* and *In Memoriam* (1850), and arguably crafted in these poems the most eloquent articulations of the complex Victorian response to geology.

Like most of his peers, Tennyson was familiar with the young science of geology and its implications for nature and time. Contemporary scholars have documented Tennyson's reading of assorted articles and books on geology, among them Lyell's *Principles*, Mantell's *Wonders of Geology*, Robert Chambers's *Vestiges of the Natural History of Creation*, and Hugh Miller's *Testimony of the Rocks*.[3] Even without reading these introductions to geology, Tennyson could hardly escape the pervasive discussions of the new science and its implications in the periodicals of the day. A regular reader of the *Quarterly Review*, Tennyson was exposed to writing about contemporary science topics: in their introduction to *Science in the Nineteenth-Century Periodical*, Dawson, Noakes, and Topham observe: "around one tenth of the articles published in the *Edinburgh* and *Quarterly* in the 1830s were devoted to scientific subjects, and other articles often broached scientific themes. Moreover gentlemen of science like David Brewster and William Whewell who wrote at length in the reviews clearly saw them as important platforms for addressing a non-specialist but culturally powerful public" (11). Indeed, the nonspecialist audience for science writing in both periodical literature and book form required the science writers themselves to adopt literary strategies to make their theories comprehensible and to quell negative reactions that may have ranged from anxiety to fierce antagonism. Like many of his generation, Tennyson was awed by the terrible implications of what he read in Lyell and Mantell, and with the new theories of geology, astronomy, and eventually evolutionary biology so widely disseminated, there was no escape from those implications.

Before turning to *The Princess*, I will briefly survey Tennyson's response to science in *In Memoriam*. This subject has been well examined; however, it is worth briefly revisiting here to set my consideration of *The Princess* in the context of the scholarship on Tennyson and science[4] and also to foreground the context in which Tennyson wrote the two poems.[5] Put very simply, *In Memoriam* illustrates Tennyson's anxious response to geology. Indeed, the poem spoke for a generation and has come down through the years as the century's most eloquent outcry against science's erasure of human significance. In this elegy on the death of traditional time and nature as much as on the death of a friend, Tennyson personifies a callous nature who cries, "So careful of the type? . . . / . . . / A thousand types are gone; / I care for nothing all shall go" (56:1–4), and faced with this indifferent nature, the poet despairs. These terrible recognitions of the true state of the Earth appear throughout the first half of the poem. In lyric 35, for instance, Tennyson writes that he hears,

> The moanings of the homeless sea,
>   The sound of streams that swift or slow
>   Draw down Aeonian hills, and sow
> The dust of continents to be;
>
> And Love would answer with a sigh,
>   "The sound of that forgetful shore
>   Will change my sweetness more and more,
> Half-dead to know that I shall die"
>   (35: 9–16)

Tennyson clearly evokes a Lyellian view of landscape shaped by water; over the "long result of time" continents erode to nothing more than dust. In that context, love, even capitalized and given voice, is nothing more than a simple sweetness forgotten by nature and already dying with the knowledge that death, of the individual as well as the species, is inevitable. The "more and more" Tennyson uses in line 15 echoes line 142 of "Locksley Hall": "The individual withers, and the world is more and more." Eleanor B. Mattes considers the poem Tennyson's reaction to the shock of discovering in Lyell that nature offers no promise of immortality for humans, yet, she argues, Tennyson takes comfort in the presumption that nature does not reveal all.

In *Tennyson and Geology*, a monograph published by the Tennyson Society, Dennis Dean meticulously analyzes the lyrics influenced by science and links each to the geology text to which Tennyson arguably responds.[6] Dean asserts that readers have been so impressed by Tennyson's apparent anticipation of Darwin's evolutionary theory that they have failed to notice his complex representation of geological debates in the elegy. Even those who have recognized Lyell's influence on the poem have often failed to consider how Tennyson engages with competing views of geology. Dean argues that the lyrics reveal "a broadly based appraisal of geology that went through several phases" (1985, 9). For example, he claims that lyrics 34 to 36 of the poem reflect the Huttonian theory of an Earth with a fiery center. Dean reports that Tennyson wrote these lyrics between 1833 and 1836, during a period in which he read widely in the geological literature but had not yet read Lyell. Evidently, he did not read *Principles of Geology* until 1836, but lyric 35 clearly shows that Tennyson, like Lyell, learned uniformitarianism from earlier geologists, like Hutton.[7]

Dean claims that Lyell's influence on *In Memoriam* first manifests itself in lyrics 43, 69, and 70, probably written between 1837 and 1839 (1985, 10). In lyric 43, for instance, Tennyson offers a comforting view of uniformity: "So that still garden of the souls / In many a figured leaf enrolls / The total world since life began" (10–12). If processes in action in the distant past can be observed in the present, the fundamental principle of uniformitarianism, then one can imagine that a simple natural object like a leaf might represent all of humanity's souls. This is an empowering image, in the spirit of the Lyell who triumphantly offers his despairing reader the observer on the margin of his frontispiece. Tennyson, too, exalts the power of the observer, also the writer, with his diction: the "leaf" (11) also references the page of the poem. Thus, as a drop of water can change the Earth, so Tennyson makes great claims for his poem. Of course, Dean also considers lyrics 55 and 56, those most often connected to Lyell's *Principles* and the alarming implications offered there. In these lyrics, Tennyson falters and fails to find any source of empowerment: "Nature, red in tooth and claw" forces the speaker to ask the awful question— Shall "man be blown about the desert dust, / or seal'd within the iron hills?" (56:19–20). He concludes, "O life as futile, then, as frail!" (24).

Yet, just as the speaker moves through stages of grief to finally accept the loss of his dear friend, so does he advance from despair to acceptance of the new scientific worldview. In lyrics 118 and 120, he resolves the crisis he articulated so famously in 56. He establishes the context in the first line of 118: "Contemplate all this work of Time," clearly evoking uniformitarianism and geological time. He then sets humans in that context, but rather than seeing them obliterated in the face of deep time, he now finds the ability to elevate human life above the "seeming-random forms" and the "seeming prey of cyclic storms" (10–11). If he internalizes the "work of Time," man will be "the herald of a higher race" (12–16). Tennyson offers here a striking contrast to "Locksley Hall," in which the individual withers as the world is more and more: here, the progressive state is transferred from the Earth to the individual, and "from more to more" (17) enhances the value of the individual rather than eroding it.[8] He goes on in the same lyric to claim "that life is not as idle ore" (20) and to describe how human life can be shaped and used. Evidently, the emotional paralysis of lyric 56 has been eased. Lyric 118 concludes with a wonderful endorsement of evolutionary theory that at once accepts the theory and rejects its belittling of humans: "Move upward, working out the beast, / And let the ape and tiger die" (27–28).[9] Tennyson clearly seems to have come to terms with science, and both Dean and Mattes credit Chambers with Tennyson's ability to reconcile himself to geology. He is able to incorporate its terms, in this case the terms of evolution, into his view of humanity without compromising his sense of the supremacy of humans. In lyric 120, he dismisses the troubling implications with a couplet that neither his speaker nor his reader could have imagined fifty lyrics before: "Let Science prove we are, and then / What matter Science unto men" (6–7). Dean explains, "In general, the geology of *In Memoriam* represents a debate between two contradictory earth voices. One of these is the cyclical, nonprogressive, and wholly materialistic geology of poem 35; the other, developed gradually throughout several poems, is linear, progressive, and teleological" (1985, 13). Dean joins with many other readers of Tennyson in seeing as central the struggle to face geology and maintain a teleological view.

In her essay on *In Memoriam*, Susan Gliserman, like Beer, considers a scientific discourse that includes nonscientists as well as nonscientific language as traditionally defined. She discusses the importance of "eighteenth-century concepts, Biblical phrases, or commonly shared imprecise ways of talking about issues" and argues, "This shared body of language frequently becomes extremely important to Tennyson in his own efforts to communicate his response to the affective meanings of the science he read" (438). Gliserman suggests an approach to Tennyson that is quite useful: she sees him as a contributor to the Victorian culture of science, and she thus elevates his reflections on science in *In Memoriam*, and by extension in *The Princess*, above the simple level of reaction. We find confirmation of Tennyson's role as an active participant in the construction of scientific discourse in the letter he wrote to Edward Moxon in 1844 requesting that Moxon acquire for him a copy of *Vestiges of the Natural History of Creation*. Tennyson writes, "it seems to contain many speculations with which I have been familiar for years, and on which I have written more than one poem" (Lang and Shannon 230). In this instance, Tennyson asserts that his reading of the relevant geological text followed his writing about some of its ideas. In this sequence we see the sort of discourse Gliserman describes. She goes on to show how Tennyson sets different sciences into communication with one another, using astronomy as a more optimistic response to geology: "To organize the nurturing cosmos of *In Memoriam*, Tennyson frequently draws on his reading in astronomy; to organize the landscapes of a hostile and aggressive environment, he draws on his reading in geology" (442). Her reading actually imagines *In Memoriam* as itself teleological, whether the sciences that informed it were or not. Unlike Dean, who reads the poem as a collection of responses to distinct voices in geology, Gliserman reads it as a deliberate and entire engagement with a complex set of scientific and cultural problems. As Gliserman explains, *In Memoriam* teaches readers how to read "the science of their culture" (445).

Gliserman attends to the mutual production of a culture of mediation: the science writer frames his theories for a reluctant public, who in turn reflects upon those theories and reshapes them as it sets these new ideas against nonscientific ways of viewing the world; the scientist then responds to the revised position of the public, even borrowing its language to reframe his science. Gliserman describes this shared nonscientific language. She explains, "Science writers often use language and phrases drawn from other contexts to make their audiences aware of the emotional meanings of what appears to be neutral information" (438). She uses as an example Tennyson's phrase from lyric 56 of *In Memoriam*, "Behind the veil, behind the veil," noting that science writers (whom Tennyson read) such as astronomer John Nichol and physician Neil Arnott used the same metaphor in similar fashion. Chapter 2 considers the science writers and their efforts to use language and narrative to assert the significance of the individual; this chapter examines Tennyson's response and his own insistence on the power of humans despite (and perhaps because of) geological time in *The Princess*.

Unlike *In Memoriam*, *The Princess* is a comedy, or "tragic-comic" as the poem's frame narrator describes it. Yet, even with its lighthearted tone, it explores both the threat and the promise of geological time. As Glen Wickens and Milton Millhauser independently argue, *The Princess* works out the conflict between the bright and dark sides of science, the side that maintains a teleology and the side that does not. Written while Tennyson was writing *In Memoriam*, *The Princess* shares with the elegy several central themes: both poems reflect the poet's angst and the angst of an age faced with the alarming insignificance of humanity, and both mourn the loss not of a single human life but of the entire species. Both, furthermore, ultimately offer comfort in the form of human love and socially constructed bonds that soothe by seeming to defeat deep time.[10] In *The Princess*, Tennyson seeks to provide a comic ending to the story of nature's tragic indifference. To do so, he crafts a medley, a form that at once models and mocks both literary and scientific collection.

"To Bind the Scattered Scheme"

Tennyson labels *The Princess* a "medley," describing the fictional composition of the poem as well as its content. The form also, I argue, alludes to both geological literature and to the Earth itself. Within the frame, seven male narrators recite the story, each taking a turn and picking up where the last storyteller left off, and the women in the party contribute "intercalary" songs interspersed throughout the poem. The poem is thus a comic literary miscellany that includes a musical element and so fits neatly within the standard definition of medley. "Medley" is also used to describe battle, and the poem includes combat. Most generally, "medley" means a combination or mixture, and the poem is certainly that, with its conjunction of serious and comic themes, its competing voices, and the mixture of contemporary themes, including both women's education and geology. In his choice of a generic label, Tennyson signaled at once several aspects of *The Princess*; in addition to the standard definitions already mentioned, I will argue that Tennyson uses the label to reference the temporal medley associated with a collection of traces, each excavated from its own historical moment and then brought together and displayed in the present, whether in a museum or a poem.

Tennyson complicates the frame narrative, weaving a complex relation between past and present that runs throughout the poem: the prologue and epilogue are both set in the present, and the recognizably immediate present of the 1840s at that, while the seven sections of the internal narrative are set in the medieval-inspired past; yet the intercalary songs that appear between each section represent intrusions of the present into the past of the narrative. The poem takes place in both the distant past and the immediate present, being thus a temporal medley, as well as a literary one.[11] The interplay between past and present, each represented physically by the distinct segments of the poem, recalls the familiar landscapes of geology, strata associated with different times layered atop one another and even igneous intrusions (rock from one period intruding into a stratum from another) being much like the intercalary songs. It might be a stretch to consider the medley an allusion to the geological view of the Earth were it not for the geological expedition of section 3, in which the main characters rely on the landscape to structure their debate about women and education.

Many readers of the poem have noted how well suited to the age the medley form is as it expresses the mixed emotions of a people captivated by progress and nostalgia. In her article on the poem's form, Eileen Tess Johnston writes:

> Its genre, style, imagery, characterization, plot and narration can all best be understood in relation to medley. Medley is the formal realization of the poem's central vision of human potentiality, both the individual's and society's, and lends itself to the celebration of those qualities Tennyson wished to affirm: variety, inclusiveness, energy, receptivity, and harmonious order. Ultimately, *The Princess* projects

towards human beings in society an attitude that is itself a medley—
a mixture of hopefulness and skepticism. (549)[12]

This mixture of hopefulness and skepticism characterizes the spirit of the Victorian age in many respects but particularly with respect to the new developments in science. John Killham also connects the form to the age, reading it as Tennyson's peculiarly bizarre, hodgepodge approach to the many facets of contemporary life. In *Tennyson and* The Princess, Killham writes of the medley form: "Certainly the bizarre medley of incidents drawn from fairy tale, science and mediaeval chivalry is in fact nothing other than a light-hearted representation of the extraordinary complexity of contemporary thought and taste" (276).[13] Both Killham and Johnston call the poem a fairy tale, and I argue that the medley is a central element of this particular fairy tale: it delights in multiplicity while simultaneously celebrating the ability to assemble the multifarious into a cogent whole. This ability is demonstrated also by geologists who assert authority over the Earth's fragmented story with their narratives: in this likeness, we can see the fictional poet-narrator of *The Princess* as akin to the geologist-writer, who is the subject of chapter 2.

Killham discusses two contemporary reviewers who noted the resonance of the medley form, considering it a reflection of the nineteenth century. In 1849, Aubrey de Vere wrote in the *Edinburgh Review*, "If a man were to structure the external features of our time, for the purpose of characterising it compendiously, he would be tempted, we suspect, to give the task up before too long and pronounce the age a Medley" (388). In de Vere's view, the form of the poem models the eclectic quality of the age; the specific elements of Tennyson's medley are less important than the nature of the form itself. One year later, Charles Kingsley observed in *Fraser's Magazine* that Tennyson "makes his 'Medley' a mirror of the nineteenth century, possessed of its own new art and science, its own new temptations and aspirations, and yet grounded on, and continually striving to reproduce, the forms and experiences of all past time" (250). Kingsley attends more concretely than de Vere to the fragments Tennyson assembles in *The Princess*. The poem conjoins sociopolitical debate with scientific concerns, and present-day displays of technological progress with past displays of valor. Each element of the medley is a trace of something greater than itself yet altered when set within this collection of traces. Like a fossil defined by its position in a display case, the intercalary songs, for example, take meaning from and give definition to the narrative sections that surround them.

Tennyson makes that act of conjoining such diverse elements one of the subjects of the poem when he describes the poet-narrator's assembly of assorted literary fragments into the whole of the poem. Although he endeavors to stay true to the original collection of narratives, this character inevitably imposes his own singular style in his re-narration. He describes the composition process in the conclusion:

So closed our tale, of which I give you all
The random scheme as wildly as it rose:
The words are mostly mine; for when we ceased
There came a minute's pause, and Walter said,
"I wish she had not yielded!" then to me,
"What if you dressed it up poetically!"
So pray'd the men, the women: I gave assent:
Yet how to bind the scatter'd scheme of seven
Together in one sheaf? What style could suit?
                                    (Conclusion: 1–9)

The poet strives to construct a unified tale from the collection of story fragments. He describes his voice as falling along a "strange diagonal" (27) in his
effort to appease both the male storytellers who prefer a comic, mock-heroic
tale and the female songstresses who "wished for something real" and pushed
the poet toward "true-heroic—true-sublime" (18–20). Many critics have found
this "strange diagonal" a useful model to understand not only the poem's tone
but also its political position. For example, Marjorie Stone and Eve Sedgwick
see the diagonal as symbolic of the breakdown of clear gender categories in the
poem. Others have considered the "strange diagonal" as Tennyson's own diag-
nosis of what is wrong with the poem. Christopher Ricks explains:

> Tennyson wrote [the poem] after the worst five years of his life, and
> he created a complicated series of evasions such as could temporarily
> stave off his dilemmas and disasters. . . . It evades the artistic prob-
> lems of the long poem, of narrative, drama, and lyric, by proffering a
> "Medley." Similarly, the moral issues of women's education are in-
> voked and evaded by a "strange diagonal." (1987, 182)

Even Ricks, however, seems to admire the skill of Tennyson's evasions, and
others see what Ricks calls evasions as clever compromises. Isolde Karen Herbert, for instance, considers the autonomy of the frame section of the poems
and argues that the tension between the frame and the interior narrative allows
the poet "to create and confront the forces of insurrection silently and unob-
trusively while remaining within the protected enclosure of the dominant poetic voice" (146). She also describes the frame as the "means to express, yet
control, revolutionary social ideas" (145), and in this dual stance, Tennyson
successfully charts a diagonal through contradictory and complicated matters.
I agree with Herbert that the poem is deliberately conflicted, and in that internal conflict—represented by the "strange diagonal"—Tennyson at least pre-
tends to resolve several of the problems that Ricks sees him seeking to evade.
The problems that interest me here are those raised by geology.
    While medley as a literary term indicates first and foremost miscellany,
the form was newly familiar to Tennyson's readership in the 1840s under a

different guise, and that is the scientific texts that had become so popular. The poet had a great model for his fragmented narrative in the Earth itself, and, more importantly, for the cohesive representation of that story in the writing of geologists who applied order and teleology to the otherwise disjointed and obscure evidence of the Earth's history. Writers like Charles Lyell and Gideon Mantell applied their expertise to the scattered array of rocks, fossils, and bones, constructing from these relics a cogent report of events that occurred long before humans could witness them. Their ability to infer such events and larger patterns conferred upon them the authority, previously wielded only by Genesis, to describe the creation of the Earth, as it is known in the present. In addition to the scattered scheme of the Earth itself, the geologists generally produced texts that were themselves medleys of explanation, moving categorically through different regions, formations, or types of fossil. The texts were literary versions of the collections found in museums and drawing rooms. Tennyson's narrator borrows the form of the geological text and the authority of the geologist to construct a whole from fragments. This narrator-poet, whose role is made clear in the conclusion, translates the disconnected strata of the storytelling into the final, monophonic narrative that is the poem.[14] That Tennyson goes to such great lengths to establish the fictional composition process of this poem confirms that the nature of the narrative is fundamental to the poem's meaning. In other words, the authority to make sense of fragments and to represent them as a whole to a readership is essential to Tennyson's project in *The Princess*, just as it was essential to the geologist-writers I discuss in chapter 2.

The poem's brief prologue establishes the nature of the medley, the authority of the narrator to ascribe cohesion to disparate objects and issues, and the peculiar conjunction of gender and geology. Tennyson assembles signs of the present time through the technological festival that serves as the poem's backdrop. Indeed, a line from the prologue, cut from the 1850 edition of the poem, implies the scene is a microcosm of the present day: "The nineteenth century gambols on the grass" (232, 1847 ed.). The reader learns in the prologue that the events of the poem are set amidst a miniature Great Exhibition, a scientific festival, taking place on the lawns of Sir Walter Vivian's home. Providing a partial source for this setting, Killham reports that a Mechanics' Institute festival was held on July 6, 1842, at Park House, near Maidstone. This was the home of Edmund Lushington who married Tennyson's sister, Cecelia, a few months after the festival, and it is likely that Tennyson attended the July festival. He certainly knew of it, and it is reasonable to imagine that he associated the festival with his sister and her marriage. Killham reprints a description of the festival, which appeared in the *Maidstone and Kentish Advertiser*. Tennyson's setting is certainly similar:

> Cricketing, trap bat and ball, and various games were played in different parts of the ground, a cannon was occasionally fired off, ignited by

a spark from an Electrical Machine at a distance of about 20 yards; at another part of the grounds a model of a steam engine was at work, turning a circular saw with great rapidity, and the model of a steam boat plying round a light house, a table spread with philosophical instruments, consisting of telescopes, microscopes, etc. etc., formed a very interesting source of amusement. (qtd. in Killham 61)

In his prologue, Tennyson also describes a cannon, a steamer, a telegraph, and telescopes (66–77). These technological objects are signs of the times and indications of progress. In the face of withering individuals, they assert human achievement, particularly in the areas of communication and scientific study. The steamer, the telegraph, and the telescope allow people to supersede great distances and to close gaps in space and time. In a fragmented world, these tools promote cohesion and lend great authority to those who wield them.[15]

In addition to hosting the Mechanics' Institute festival, the Vivian estate houses a private museum of sorts, an eclectic display of objects collected from all over the world and from many different times.[16] Primarily, the museum contains geological and what would later be considered anthropological artifacts, the fruits of industrious excavation. The museum is likely modeled on the Mantellian museum, Gideon Mantell's private collection of artifacts open to visitors in the late 1830s. While there is no record of Tennyson visiting the museum, its broad repute ensures that Tennyson knew of the museum, and given the museum's large number of celebrated visitors, it is certainly possible that Tennyson was among them. In any case, Mantell's son, Walter, often acted as curator for his father's museum, and Tennyson may allude to the Mantellian museum when he names Sir Vivian's son:

And me that morning Walter show'd the house,
Greek, set with busts: from vases in the hall
Flowers of all heavens, and lovelier than their names,
Grew side by side; and on the pavement lay
Carved stones of the Abbey-ruin in the park,
Huge Ammonites, and the first bones of Time;
And on the tables every clime and age
Jumbled together; celts and calumets,
Claymore and snowshoe, toys in lava, fans
Of ancient sandal, amber, ancient rosaries,
Laborious orient ivory sphere in sphere,
The cursed Malayan crease, and battleclubs
From the isles of palm: and higher on the walls,
Betwixt the monstrous horns of elk and deer,
His own forefathers' arms and armour hung.
(Prologue: 10–24)

The description of the house presents the whole poem in miniature: objects that suggest the domestic—toys and sandals—are ranged alongside objects of war—the crease and battleclubs. Past and present are displayed side by side, and thus, the museum, like the poem, shows an effort to control time. The museum even includes geological relics; epic frames epoch, the ammonites on the patio indicating the poet's equation of the narrative of deep time with the domestic and heroic narratives. This collection of diverse artifacts, itself a temporal and visual medley, frames the poem's narrative and foregrounds the empowered positions of both collector (poet) and visitor (reader). It also reminds the reader that history "is a knowledge by traces" (Ricoeur 1985, 120). This museum and this poem are built of traces that, in being assembled, tell a story. The medley of traces, of topics, of styles, of voices, and of times offers the reader a rich array of fragments to examine alongside one another, and the frame-poet acts as the curator bringing some sense to the array. The prologue's medley of telescopes and fossils surprisingly introduces Tennyson's foray into the murky matter of women's education. Indeed, the prologue teaches the reader to consider the woman question alongside the telescope and the ammonite; all these traces reflect the Victorian age. In the next section of this chapter, I will examine how Princess Ida is herself both a geologist and an object for geological study.

Ida Red in Tooth and Claw

Tennyson's representation of nature as "red in tooth and claw" reflects his interpretation of a world governed by geological time. In such a world, human lives are utterly insignificant and, while there may be order in the universe, that order has nothing whatsoever to do with humanity. Time's expanse belittles people so dramatically that, for all intents and purposes, they cease to exist—they "wither" away to nothing. Deep time's nature does not nurture individual lives or even particular species; this nature demonstrates total indifference to men and women, attending only to processes that are somehow extra-temporal and the resultant changes that take a million lifetimes to effect. It is perhaps not entirely right to conflate time and nature, but the nature that so troubles Tennyson is the nature born out of the expanded time scale. Furthermore, the signs of time were read in the geological landscape, so the conflation came easily to the Victorians. Nature only becomes "careless of the type" when it expands to fill the millions and millions of years of geological time. Rather than denying this expanse of time and its frighteningly careless nature, Tennyson works in his poetry to describe a way of living in this new world. In the balance of shadow and substance, the guise of something and the void of everything, *The Princess* offers the possibility of meaningful life lived in deep time.

This possibility depends on the peculiar conflation of two seemingly unrelated elements of mid-Victorian life: family and geology. Some critics have focused on only one of these two elements of the poem, ignoring the other; others have dismissed the odd conjunction as unrelated topics of a medley intended to provide an overview of the age. I argue instead that the family and geology axes of the poem, just like the past and present or the comic and tragic sections, are essential to the medley's "strange diagonal." Tennyson wrote *The Princess* when the movement for women's education was gaining strength. His friend F. D. Maurice began work in 1843 on Queen's College, initially designed to certify governesses, and some readers have considered Ida's university a parody of Maurice's.[17] In his notes on the poem, Christopher Ricks suggests as a possible source Hannah More's *Female Education* (1799), which was in Tennyson's library, and Ricks also notes several articles on the subject of women's education that appeared in encyclopedias and in periodicals. Killham associates Tennyson's Ida with the Saint-Simonians, who "believed that woman should be granted independence and education in order to share in the progress of the new society based upon universal co-operation" (32). Yet, the desire of many women to pursue higher education threatened traditional ideals of women as devoted exclusively to their families. These women denied the dictates of the biological and cultural imperatives to nurture, or so many middle-class men and women believed. Opponents of women's education believed that the debate itself, like the discourse surrounding geology, called into question the feminine nature.[18]

In an article that considers the relationship between the nature of woman and the nature of nature in a mid-Victorian world heavily influenced by science, James Eli Adams deconstructs the Victorian notion of Mother Nature. He connects Tennyson's concerns about geology to cultural anxiety about women's role in society: "Tennyson's poetry, like Darwin's science, reminds us that the maternal ideal assumes its urgency in Victorian culture at the same moment that the claims of science are finally undermining one of the culture's most powerful icons of maternity: the very conception of nature as Mother Nature" (9). Considering the personification of nature in *In Memoriam*, Adams argues that geology inspired the chilling idea that, in the glaring light of modern science, "'Mother Nature' [had] become a savage oxymoron" (7). In other words, "Nature" became antithetical to the ideal of "Mother." Simultaneously, that ideal was challenged by science and by a society that began to ask if women's role should be so prescribed. Adams asserts, "to question the nature of 'Nature' . . . is inescapably to question the nature of woman" (8). Written concurrently with *In Memoriam*, *The Princess* also examines savage nature and wonders, in turn, about savage woman.

Whereas Tennyson cringes before "nature red in tooth and claw" in *In Memoriam*, he writes a romance about living with deep time in *The Princess*. In this medley of mock-epic romance and social commentary, Tennyson allies Princess Ida's feminist project with cruel nature, but he ultimately subdues her feminism and describes her return to conventional domesticity. Ida threatens familial, political, and natural order when she flouts a legal marriage contract and instead forms a women's university that completely excludes men and, by extension, traditional family. Throughout most of the poem, she exhibits characteristics unexpected from a woman in the mid-nineteenth century, or in the abstractly medieval time in which the fictional princess lives: she is fiercely intellectual and positivistic, and, more troubling to a Victorian readership, she spurns family and adopts a callous death penalty for male intruders in her educational utopia. In her indifference to men and to motherhood, Ida embodies the fearful nature Tennyson perceived in a world inflected by geological time. She explicitly embraces the expanded time scale in section 3 of the poem when she leads a geological field trip and expresses regret that women's lives cannot be lived on that scale: "Would, indeed, we had been, / In lieu of many mortal flies, a race / Of giants living, each, a thousand years" (3:250–52). Yet, Tennyson reasserts a familial narrative over the infinite expanse of time and its cruelly indifferent Mother Nature with the dissolution of the university and Ida's concession to marriage. In contrast to the "fairy tales of science and the long result of time," *The Princess* offers a fairy tale of human importance and the centrality of the domestic narrative.

In Tennyson's poem, artifacts, genres, and times converge to tell a tale that satisfies the troubled longings of an age whose world no longer moves along a familiar narrative path. The frame introduces the poem's strange conjunction of gender and geology, and Ida's assistant Psyche stresses this conjunction in her lecture on women's history, in which she explicitly links geology with proto-feminism. She begins her history in a distant past, akin to that described in Genesis: "'This world was once a fluid haze of light, / Till toward the center set the starry tides, / And eddied into suns, that wheeling cast / The planets . . .'" (2:101–4). Tennyson acknowledges that Psyche describes "the nebular theory as formulated by Laplace" (qtd. in Ricks 1987, 209). Pierre-Simon Laplace (1749–1827) was an eighteenth-century astronomer who offered the nebular hypothesis to explain the creation of the planets; he is credited with conceiving one of the first theories of evolution, though his theory applies to the heavens and not to the Earth or its inhabitants. Loren Eiseley lists Laplace among the many thinkers who paved the intellectual way for Darwin's theory of evolution, but Eiseley is careful to explain that the implications of Laplace's hypothesis for the Earth were not realized until the nineteenth century when it joined forces with many others to reveal the depth of Earth history and the possibility of species evolution. Eiseley writes:

> Laplace had been content, toward the end of the eighteenth century, to propose his nebular hypothesis as to how the planets might have been formed. That this in its turn suggested long lapses of astronomical time there can be no doubt. Still, Laplace did not ask of the heavens the questions the nineteenth century was to ask; he did not debate the secular cooling of the earth or the rate at which the sun was consuming its own substance. (330)

Eiseley locates Laplace firmly in an eighteenth-century context, and while Tennyson explicitly references Laplace, we must remember that he writes in the nineteenth century. The questions that Laplace did not ask of the heavens or, more importantly, the Earth, had become urgent questions for Tennyson and his contemporaries. Milton Millhauser argues in a note on *The Princess* and Chambers's *Vestiges of the Natural History of Creation* that Tennyson probably takes the nebular hypothesis not directly from Laplace but rather from Chambers, who had recently made the extension of this hypothesis to human life uncomfortably clear (1969, 22).

Psyche uses the nebular hypothesis, with its single act of creation (rather than the series of creative acts as recounted in Genesis) to emphasize the power of the feminine in the early formation of the universe.[19] She mentions no God, only the shaping power of the tides. With this theory of the universe, Psyche simultaneously evokes uniformitarianism and genders uniformity feminine. Tides, linked to lunar cycles, are associated with women, and in Psyche's history, the moons and their tides precede the suns: women come before men. In addition to thus rejecting the biblical creation story and the subsequent tale of Adam's rib, Psyche also attributes creative power to the uniform action of water—ebbing, flowing, and eddying into the known universe—rather than to any higher being. She goes on to describe the evolution, or devolution in her view, of man, whom she associates with prehistoric beasts: "'. . . then the monster, then the man'" (2:104). She concludes her historical overview with primitive man who "'crush[es] down his mate'" (2:106). In this verse-paragraph, Psyche uses seven lines, paralleling the seven days of Genesis, to consolidate all of creation, but the genesis she recounts originates with women and water. While nodding to the Bible's version of creation with her form, Psyche simultaneously divides Earth history into five epochs: the "fluid haze," the development of the planets, the age of the monster, the age of man, and her own age. Here again, we can see Tennyson following Chambers, this time emphasizing the notion of succession.[20] Psyche lists expansive periods of natural history, and the implication is that the age of man has passed. Women have succeeded men as the dominant species, with Ida as their champion. With her endorsement of uniformitarianism at the beginning of this history, Psyche supplants the biblical creation narrative with the originary narratives suggested at once by geology and her particular brand of feminism.

The body of Psyche's lecture denounces sexist practices and celebrates notable women, chiefly Ida. From the "legendary Amazon" (2:110) to Elizabeth I (146), Psyche describes women who have "vied with any man" in government, war, and grace. These singular women are like inverse traces: they signal what will come, and their significance is magnified by the passage of time. She distinguishes these women from the majority who live under oppression: she claims that the woman's state in the Persian, Grecian, and Roman societies was "far from just" (116), and she scorns Salique laws, Chinese foot-binding, and Islamic practices (117–18). She considers chivalry a turning point because "some respect, however slight, was paid / To woman" (120–21), and she looks to education to offer women more than slight respect. Ida's university promises reform for women the world over, and Psyche acknowledges women's deep "debt of thanks to her who first had dared / To leap the rotten pales of prejudice" (125–26) and enable women to "disyoke their necks from custom" (127). Psyche mentions a common objection to educating women and insists that if women do have smaller brains than men, their brains, like hands, will grow with use (131–35). She thus implies that female brains have been stunted by custom. As Psyche presents it, the university rests at the apex of women's history and offers women hope of intellectual liberation in its challenge to social norms.

The head of the university, Ida represents what many opponents of women's education most feared: accepting education meant rejecting marriage and motherhood. For Tennyson, these concerns blend with more abstract questions about the nature of nature, and what it means when nature seems to have no relation to human families and individual lives. The central conflict in *The Princess* is not the existence of the university. Ida established the institution peacefully with the support of her father, and, when the story begins, the university has already existed long enough to suffer internal political turmoil (the transfer of power from Blanche to Psyche). Rather, Ida's rejection of the marriage contract causes the conflict. The prince reports that despite his having been "proxy-wedded" to the princess in his youth (1:33), Ida renounces the commitment. Her father "said there was a compact; that was true: / But then she had a will . . . / . . . / . . . certain, would not wed" (46–49). Ida's refusal to marry troubles the men both in the poem and in the frame, but perhaps even more troubling is her failure to honor the "compact." The compact is a legal obligation, but it also represents society's standard and expectation. Young girls are promised to a marriage market, and the presumption is that they will dutifully marry and produce children, whom they will love and nurture, thus propagating not only the species but also the cultural standard. Ida's disregard for this standard evokes disconcerting fears that the standard is a mere construct. Ida certainly voices the mid-Victorian feminist insistence on a woman's right to education, but she also personifies nature, recently understood to be a lawless, even cruel force. The pressing need to tame her results, at least in part, from the anxiety inspired by geology.

The frame characters determine to create a woman who is "Grand, epic, homicidal" (Prologue: 219). Indeed, Ida threatens not just individual men, but all men. The rules of her university exclude men and even prescribe capital punishment for any who dare to intrude. Florian asks with incredulity, ". . . who could think / The softer Adams of your Academe / O sister . . . were such / As chanted on the blanching bones of men?" (2:179–82). His phrasing evokes the fossilized bones of a prehistoric animal and the prehistoric humans who undeniably lived in the distant past, and he reminds the reader that the terms of *The Princess* transcend the individuals of the story to include the whole of the human species. Ida seems very like the cruel nature of *In Memoriam*, "so careless of the single life" (55:8). The prince's father associates her with violent forces of nature: he vows to ". . . send a hundred thousand men, / And bring her in a whirlwind" (1:63–64). Scores of soldiers and a natural disaster are required to tame Ida's "Wild nature" (5:165). When Ida rails against this violent penetration of her university, her "cheek and bosom brake the wrathful bloom / As of some fire against a stormy cloud" (4:364–65). The power of fire and storm supplant the soft cheek and tender bosom of womanhood. The prince attempts to assuage her ire by recollecting his youthful fantasies of a kind and gentle future bride. She responds with rage in the spirit of a catastrophic geologic event: " . . . a tide of fierce / Invective seemed to wait behind her lips, / As waits a river level with the dam / Ready to burst and flood the world with foam" (4:450–53). A dam is literally a construct, a structure designed to contain water and restrict its movement. Here, as in Psyche's lecture, Tennyson connects female strength to water's potency. Tennyson explicitly likens Ida's rage, and possibly Ida herself, to a great flood, a flood that would destroy man's construct and shatter life as we know it.

Not only does Ida demonstrate an appalling indifference to men and instruct her female students to follow her rejection of men and conventional domesticity, but she also scorns motherhood. We see this cold carelessness at its extreme in section 4, when Ida callously proposes to leave Psyche's infant to its death: ". . . she called / For Psyche's child to cast it out from the doors" (4:218–19), where, abandoned and alone, it would surely die. Significantly, Ida does not try to murder the baby; rather, like nature, she simply and cruelly rejects it. Ida's lack of maternal instinct and the implication that others share her views horrified many contemporary readers. Both Kingsley and de Vere commented on the "unsexing" of women in their reviews of the poem. Kingsley wrote:

> In every age women have been tempted . . . to deny their own womanhood . . . Cleopatra and St. Hedwiga, Madame de Staël and the Princess, are merely different manifestations of the same self-willed and proud longing of woman to unsex herself. . . . [Tennyson] shows us the woman, when she takes her stand on the false masculine ground of intellect, working out her own moral punishment, by destroying in herself the tender heart of flesh. (1850, 250)

De Vere comments even more specifically on the issue of children in the poem, writing: "Many passages in it have a remarkable reference to children. They sound like a perpetual child-protest against Ida's Amazonian philosophy . . ." (400). Indeed, the poem is a "perpetual child-protest" of the Earth's children against their neglectful and distant mother, Nature. The figuration of an alien nature in woman draws on the same angst provoked by the recognition of geological time.

Ida and Uniformitarianism

Ida, "grand, epic, homicidal" indeed, demonstrates the cruelty and indifference of nature as it was understood in the light of geology. Tennyson even
links Ida to time itself. When the prince describes how Ida has inflected the
landscape of his life—"at eve and dawn / With Ida, Ida, Ida, rang the woods"
(4:412–13)—he echoes the famous words of George Poulett Scrope who
wrote, in 1827: "The leading idea which is present in all our researches, and
which accompanies every fresh observation, the sound to which the ear of the
student of Nature seems continually echoed from every part of her works,
is—Time! Time! Time!"[21] The thrice-repeated "Ida" suggests Scrope's similar construction and draws a parallel between Ida and time. Tennyson affirms
the connection later in section 4, when Ida herself extols deep time. She
preaches to her students, "Six thousand years of fear have made you that /
From which I would redeem you" (486–87), referring explicitly to the biblical chronology of the Earth and associating this false history with conventional domestic roles. In explicitly connecting women's position in society to
the biblical time scale, Ida suggests that women have been forced into a social
position that is itself a false trace—that is, a remnant of a world view science
assures us is wrong. Ida's proposed redemption of women has to do with intellectual empowerment and also with acceptance of geological time. In the
vastness of geological epochs, women's identity need not be linked to the
trace of Adam's rib; rather, women might leave their own mark on time. Tennyson most obviously aligns Ida with deep time in section 3, where a geological expedition provides the occasion for Ida and the prince to debate her
commitment to the university and to ameliorating women's position. Against
the sublime backdrop of fossils, bones, and rivers that rage through jagged
cataracts, Ida positions herself in the uniformitarian versus catastrophist controversy. She exploits the geological imagery to assert her hopes for women's
progress over time.

    Revealing Tennyson's familiarity with geology and its debates, the expe-
dition scene includes both the catastrophist and the uniformitarian interpretations of geological traces. Ultimately, however, this episode in the poem
reveals Ida's commitment to gradualism and uniform processes. She expresses
longing for a catastrophic event, which event Tennyson signals with a list of
rocks: the members of the party "chatter . . . stony names / Of shale and horn-
blende, rag and trap and tuff, / Amygdaloid and trachyte" (3:343–45). Four of
these stones are volcanic in origin, suggesting catastrophe, and Ida wishes she
could explode convention with volcanic violence, reforming women's lot in a
single earth-shattering gesture: "Oh if our end were less achievable / By slow
approaches, than by single act / Of immolation" (266–68). To her chagrin, she
recognizes that revolutionary change must come through the slow processes
of uniformitarianism. Despite her desire for a great quake that will break

down the constraints on women's intellectual freedom, Ida maintains a constant faith in uniformity throughout the poem. In fact, the geological expedition reveals the fundamental principle of geology that ultimately makes Ida a sympathetic character. She does not neglect children and disregard marriage contracts because she is cruel but because her perspective is much broader than the scope of individual lives, and here Tennyson reminds us that nature is not utterly indifferent to humanity. Nature may, after all, be "careful of the type," and so is Ida. Her lack of interest in individuals is subordinate to her great concern for her sex. She regrets the tiny span of a human life and imagines a way of living in deep time that would privilege both the individual and the species:

> Would, indeed, we had been,
> In lieu of many mortal flies, a race
> Of giants living, each, a thousand years,
> That we might see our own work out, and watch
> The sandy footprint harden into stone.
>   (250–54)

Ida wishes that she could live to see the fossilization of her own work, but with her conditional tense—"would . . . we had been"—she sadly acknowledges the expanse of geological time and regrets that social change must be "the long result of time." The sandy footprints left by her generation will be traces for the women of the distant future to excavate and narrate, a narration offered in the poem by the women of the frame.

Ida's efforts and those of her comrades leave traces; they will, in time, offer women a better life. No catastrophe will come with revolution in its wake; instead, women must look to the power of a single drop of water to wear down a mighty stone over many years. Ida lets fall such a drop—a teardrop juxtaposed with the roaring river: "She bowed as if to veil a noble tear; / And up we came to where the river sloped / To plunge in cataract, shattering on black blocks / A breadth of thunder" (272–75). These lines imply that Ida's passionate tear will over time wield the force of the river to carve out a cataract in traditional ideals of gender. Thus, while the princess's ideas may strike some readers as catastrophic, her plan is not to incite revolution but rather to work gradual change over many years. In geological time lies the possibility of social reform: the tear, as it carves over time, rages with a "breadth of thunder." Change is regretfully slow in coming, but come it will. Ida endorses and even embodies uniformitarian nature—gradual yet fiercely powerful.

Troubled by Ida's disregard of the domestic, the prince argues that, in favoring the species and rejecting marriage, Ida rejects also a woman's best accomplishment:

might I dread that you,
With only Fame for spouse and your great deeds
For issue, yet may live in vain, and miss,
Meanwhile, what every woman counts her due,
Love, children, happiness?

(225–29)

Ida does not count the traditional woman's due as her own. Instead, she places deeds above children, arguing that "great deeds cannot die" (236–37). The prince and Ida argue here over what sort of trace a woman should leave behind: revolutionary acts or children? Ida speaks here with the voice of nature, who discards individuals daily but does leave lasting monuments—"great deeds"—on the Earth's surface: how can Mother Nature be expected to worry about children when she has mountains and rivers to raise? How could Ida think of marriage when she should provide for her whole gender? Ida looks into the riverbed at "the bones of some vast bulk that lived and roared / Before man was" (277–78) and makes the crucial analogy: "'As these rude bones to us, are we to her / That will be'" (279–80). Her concern is not with the next generation but with generations to come. She looks not to her own reproductive capacity but to the intellectual capacity of an institution like her university. James Eli Adams notes, "one must appreciate that Princess Ida is a feminist of a peculiarly Victorian sort: she conceived of her goals and of social progress in generally evolutionary terms" (19). What the prince and readers like de Vere and Kingsley do not appreciate is that where Ida fails to feel a maternal commitment to her potential children, she feels a fierce commitment to the future of her sex and the uniformitarian change required to ameliorate women's position. Ida sees herself as a citizen of deep time, and, though she throws off the marriage bond, she honors the contract of uniformity.

Ida embraces geological time and deconstructs Genesis in her own creation narrative. She describes a time so vast that humans cannot truly perceive it, and she emphasizes the centrality of ongoing process in creation:

Let there be light and there was light: 'tis so:
For was, and is, and will be, are but is;
And all creation is one act at once,
The birth of light: but we that are not all,
As parts, can see but parts, now this, now that,
And live, perforce, from thought to thought, and make
One act a phantom of succession: thus
Our weakness somehow shapes the shadow, Time.

(306–13)

Ida begins with a clear allusion to Genesis; her biblical language evokes divine creation and leaves no doubt about which narrative Ida is rewriting. She offers

the delightfully abstract, "For was, and is, and will be, are but is." Ida's five copulas represent the range of possible tenses, beginning with the past, shifting to
the present, then the future, and finally back to the present. The final two verbs
in the present tense enact a shift in number. These last two are in fact the verbs
of the sentence, while the first three copulas form a noun phrase. Paraphrased,
the sentence states the essence of the past, the essence of the present, and the
essence of the future all exist in the present tense. Ida secures her claim even
further when she shifts from "are" to "is" at the end of the line. Not only do
multiple temporalities coexist, they also coexist singularly in a coeval collision
of all times. Past, present, and future are one thing, requiring only a singular,
present-tense copula. This point is reasserted in the next line: "all creation is
one act at once." The act of creation is the process witnessed by geology: we
know the past because we observe processes in the present. Thus, Ida offers as
her creation story a perfect definition of uniformitarianism.[22]

Yet, Ida proceeds to describe humans' limited ability to perceive time. She
explains that we "are not all" and "can see but parts." With her modifier to
"parts": "now this, now that," Ida suggests an imperfect view of time; we see
only fragments of the whole. "This" and "that" each occur singly as separate
events. We "make / One act a phantom of succession." We cannot perceive
the temporal interplay that the play of copulas above illustrates, nor the profound depth of time the interplay evinces; we see only a minute fraction of history stretching a short distance back and a short distance forward. She further
emphasizes her point by invoking Plato's cave parable. She describes the one
act of creation as "the birth of light" and explains that humans cannot see the
light; they see only parts. She ends her definition with the sad suggestion that
"Our weakness somehow shapes the shadow, Time." The light is deep time,
and the shadow is the construct of contemporary life. To depend on the belief
that each age exists only in its own moment is one of humanity's great weak-
nesses. To affect real change, people must reach beyond their limited perspectives of time, something Ida strives to do. It is worthwhile to take a moment to
connect Ida's description of time to the trace as described by Ricoeur and Derrida.[23] Ida suggests that all time exists at once in the present. She does not
mention a trace or an artifact, but when she turns to a person's limited view of
time, she might be referring to the encounter with time when one faces a particular moment in history or a vestige of that moment. In such an encounter,
the observer faces a fragment of time, yet if one reads that fragment as a trace,
then one also sees in the fragment the many times leading up to the construction of the trace as well as the many times that have passed since the trace's
moment yet through which the trace has endured. Ida seems to challenge the
notion of succession as weak: seeing each age as discrete is limiting. The temporal conflation of the conjoined copulas instead offers the multi-temporal
power of signification central to the trace, a power Ida seeks.

Throughout the poem, shadow represents the construct of time and the
associated social constructions of gender and domesticity. Ida blames human

weakness for shaping the "shadow, Time," and she seeks a way of living in substantial time. Surprisingly, it is her betrothed who she so resists that offers the possibility of such a life. The prince presents a complex example of living within the construct whilst perceiving the reality of time's expanse. His so-called "weird seizures" (1:14) are periods during which he cannot distinguish these two worlds. He explains, "none of all our blood should know / The shadow from the substance" (8–9). While, at first, this handicap suggests a tendency to confuse reality, it actually indicates an ability to perceive the real that lies "behind the veil" of society. The weird seizures were added to the poem in 1850, and serve to highlight not the disjointed nature of medley, as some have suggested, but rather its delicate harmony. Jerome Hamilton Buckley defends the addition of the seizures: "Far from being an unnecessary intrusion in the medley, the 'weird seizures' thus reinforce its deepest theme, the clash between shadow and substance, illusion and truth, the ultimate relation of art and life" (1960, 101). I go further than Buckley to argue that Tennyson uses the seizures not to show the clash between shadow and substance, but instead to unify them. If constructed time, the clock-time that rules human life, is a shadow as Ida claims, then the prince, with his conflation of shadow and substance, has the capacity to see the full scope of time, its quotidian construct and its geological expanse at once. Thus, Tennyson suggests, the prince embodies compromise: in marrying him, Ida can attend to the broad scope of the species while living out her own life within the domestic frame her society provides. However, Ida is not easily persuaded.

The Domestic Diagonal

The first woman to give up the university and return to familial life is Psyche. When she talks alone with the new students after her women's history lecture, Psyche discovers they are men in disguise and that her brother Florian is among them. Initially unmoved by her brother, she stands by her oath to prosecute intruders. She calls Florian "wretched boy" (2:176) and mercilessly invokes the death penalty: "my vow / Binds me to speak" (184–85). Yet, despite her first response to the male intruders, Psyche relents easily. Of course, of all the women, Psyche is the one most closely tied to the domestic as she is a mother, though not a wife. The men sway her with familiar narratives of home and childhood: "'Are you that Psyche,' Florian added; 'she / With whom I sang about the morning hills, / Flung ball, flew kite . . .'" (228–30). Inspired by such memories, Psyche quickly reverses her position and succumbs to the men's persuasion. At first, she expresses reluctance, describing herself "like some wild creature newly-caged" (281), but shortly, she gives in entirely: "It was duty spoke, not I" (288). This admission releases a floodgate of fond memories, and Psyche and Florian indulge in happy reminiscence: ". . . betwixt them blossomed up / From out a common vein of memory / Sweet household talk, and phrases of the hearth . . ." (292–94). Simply by invoking a powerful sense of the domestic, the men easily convince Psyche to give up her vow and her commitment to an inviolable intellectual space for women.

Perhaps inspired by the method Florian used to convert Psyche, the prince endeavors to sway Ida with his fond boyhood memories of her kindness. Although this initial effort fails, the Prince prefigures Ida's transformation when he presents her in sweet, soothing, natural terms. Nature red in tooth and claw gives way to a gentle, pastoral nature:

. . . when a boy, you stooped to me
From all high places, lived in all fair lights,
Came in long breezes rapt from inmost south
And blown to inmost north; at eve and dawn
With Ida, Ida, Ida, rang the woods;
The leader wildswan in among the stars
Would clang it, and lapt in wreaths of glowworm light
The mellow breaker murmured Ida.

(4: 409–16)

The prince hardly describes a person here.[24] Rather, he describes a kind nature, a nature that looks after its own and would never drown a friend, let alone a species. In this passage we see clearly that the speaker seeks to reimagine nature. The prince rewrites the "fire against a stormy cloud" (4:365) of just a few lines before as "wreaths of glowworm light," and he tries to stem the "invective. . . / ready . . . to flood the world with foam" (450–53) by relocating Ida in the murmurs of "the mellow breaker." His imagery tempers the savage nature, but his assertion of this gentle narrative proves to be only a false hope. Ida responds in anger. She does not relinquish her power, nor does she accept the imagery of domesticated nature.

Immediately upon Ida's rejection of the prince's plea, battle sounds interrupt their discourse. While the battle's objective is to force the princess into her proper position, one could argue that she never really escaped domestic bonds. Katherine Frank and Steve Dillon point out that the boundaries of the university are violated before any men physically enter the space. First, Ida's father provided the location of the university, as she could not acquire land on her own; thus the project is dependent on men from its inception. I would add that because the castle donated by the king is a summer home, the donation confers upon the university the insulting aura of a holiday lark, or a "maiden fancy" (1:48). The men in Ida's family permit her project, but they show no respect for it and, in the end, they control it. Ida's utopia has female soldiers and *The Princess* begins with discussion of a woman warrior, but men wage the battle of section 5. The prince's father, Gama, and the princess's brother, Arac, struggle to control Ida's will. Gama would have Ida succumb to her betrothed, while Arac would let his sister continue her university.[25] In either case, men seek to determine Ida's actions, just as they hope to suppress nature, red in tooth and claw. Both armies fight against gender inversion and maternal neglect. The loud clash of arms and the violent orchestration of the melee (a variant of "medley") supplant the indifference of geology's nature. As these men struggle against Ida's unconventional institution, humanity struggles against a frightening nature that defies convention.

Furthermore, Frank and Dillon remind us that men tell this tale. The seven main segments of the medley originate with the male members of the storytelling party, and the first-person speaker in each of those sections is the prince. Some women, notably Lilia, are given voice in the prologue and conclusion, but as the frame-poet crafted those sections as well, even those brief female expressions come through the voice of the male poet. The same may be said of the important lyrics that appear in sections 4 and 7: "Tears, Idle Tears," which meditates on individual loss and memory set against the backdrop of the geological landscape, is sung by one of Ida's attendants, but there is nothing in the text to indicate that the lyric originated with the women of the party; likewise, the princess sings her own song to the wounded prince in 7, but this song, too, is presented through the narrating voice of the frame-poet. In her essay on gender inversion in the poem, Marjorie Stone argues that it is appropriate to consider the passages ascribed to female characters as in a female voice, despite the intervention of the poet. She writes:

> And as for the assumption that the narrative is male, Princess Ida narrates the grandest story of all the poem—the story of evolution. Only if we focus on the Victorian frame of *The Princess* can we assume a straightforward alignment between gender and genre, story as male and song as female. Ironically, the inset story continually questions that alignment. (107)

The gender-genre association is most obviously disrupted by the prince's song, "O, Swallow, Swallow," which he sings in section 4 as a response to "Tears, Idle Tears." It is a conventional love lyric (for which he is criticized by his audience) sung in this case by a woman, for the prince is still in disguise, but with the first-person speaker being a man. Stone makes a useful point: Tennyson clearly plays with the gender of his assorted speakers, and so we should not treat the frame-poet as an absolutely singular and male voice. However, because that poet's act of pulling together fragments and crafting from them a unified narrative is given so much attention, we cannot forget that it is his voice that we hear throughout most of the poem.

The only clear exception is the intercalary songs, with which the frame-poet does not seem to have tampered. They intrude into the narrative of the poem, generally offering a gloss on its central themes. The first two present disrupted domestic scenes: a quarreling husband and wife reunited by the lasting grief over the loss of a child in years past, and a mother soothing an uneasy child whose father is away at sea. In the lyrics of each of these songs, the female singers express a melancholy sadness, rather than any sense of jubilation, at the disruption of the home scene. In other words, Tennyson suggests with these first two intercalary songs that the women in the party do not wholly endorse Ida's rejection of the marriage contract. The third intercalary song is a companion to the geological expedition of section 3; like *The Princess* itself it marvels at the expanse of geological time yet simultaneously imagines human significance in that expanse.[26] I quote from the first and third stanzas:

> The splendour falls on castle walls
>   And Snowy summits old in story:
> The long light shakes across the lakes,
>   And the wild cataract leaps in glory.
> Blow, bugle, blow, set the wild echoes flying,
> Blow, bugle; answer, echoes, dying, dying, dying.
>
> O love, they die in yon rich sky,
>   They faint on hill or field or river:
> Our echoes roll from soul to soul,
>   And grow for ever and for ever.
> Blow, bugle, blow, set the wild echoes flying,
> And answer, echoes, answer, dying, dying, dying.
>   (1–6, 13–18)

The "Snowy summits" and "wild cataract" are the ancient, immovable features of the landscape against which the sound of the bugle echoes, and in the repeated refrain, the echo sounds yet then is heard "dying, dying, dying." Tennyson's use of the present participle suggests the continuation of the sound rather than its absolute death; thus, in his diction he at once describes the end of the human impact on this sublime landscape and its continued significance—Ricoeur's trace again comes to mind. This is precisely the ambiguity Ida embraces, and this intercalary song demonstrates that the view of geology offered in the poem is shared by the female members of the contemporary party. The next two songs reflect first on a soldier driven in battle by love for his wife at home and second on the family left behind after the death of a soldier in battle. These lyrics call attention to the connection between Ida and her women and the men fighting for them. The final intercalary song returns to the conjunction of love and the geological landscape: "I strove against the stream and all in vain: / Let the great river take me to the main" (12–13). This lyric reminds the reader that the domestic ideal ultimately endorsed by the poem is set in a context that transcends human life and inevitably looks to the future and the "long result of time."

Although Ida accepts her brother as her representative in the melee (just as the women of the frame accept their male friends as their representatives in the medley), she keeps her eye on a future time when women will fight their own battles. She writes in a letter to her brother:

We plant a solid foot into the Time,
And mould a generation strong to move
With claim on claim from right to right, till she
Whose name is yoked with children's, know herself;
And Knowledge in our own land make her free.
                                        (5: 405–9)

With these lines, Tennyson refers to the fossil footprints that featured prominently in the geological discussions of the 1840s and, of course, to the sandy footprint of section 3; again, Ida sees herself and her women as analogous to the "rude bones" of the remote past. The footprint hardens over time, leaving a permanent mark for men and women of the future to study. Their study may lead them, as geology does, to know not only the past but also to understand better the present and even the future. The foot planted in time grows into future generations; indeed, Ida claims to "mould a [future] generation" with her reform efforts. The woman of the future will be free to "know herself," independently of men or children. For the immediate future, Ida must accept her brother as her best ally, but for the distant future, she marshals the support of time. She describes the inevitable prospect of uniformitarian change, moving slowly "with claim on claim from right to right."

Yet, Ida undercuts her own claims when she modulates her view toward motherhood in her letter's postscript. When she muses on her surprising affection for Psyche's baby, she yokes her name to a child's. Ida, who had vehemently renounced motherhood and maternal affection, ends her feminist treatise with a description of her bond with this child:

. . . indeed I think
Our chiefest comfort is the little child
Of one unworthy mother; which she left:
She shall not have it back: the child shall grow
To prize the authentic mother of her mind.
I took it for an hour in mine own bed
This morning: there the tender orphan hands
Felt at my heart, and seemed to charm from thence
The wrath I nursed against the world.
                                        (5: 419–27)

Tennyson enacts an important shift here. Earlier in the poem, Psyche represented motherhood, yet, now, Ida reveals an intimacy with the baby and offers herself as its "authentic mother." She characterizes Psyche as a cruel mother who abandoned her helpless infant, and she casts herself as the benevolent caregiver who will rear the child with love. In this postscript, the princess finally tempers her role as indifferent nature and adopts the role of nurturer. Her contact with the tender child initiates a nurturing relationship between nature and an individual. Ida turns away from geological time to attend to the present. Glossing the assertion of adopted maternal love, the intercalary song between sections 5 and 6 presents a melancholy tale of a woman stunned by grief when her husband is brought home dead from battle. She is numb, unable or unwilling to speak or weep; it is the child's nurse who cries, "'Sweet my child, I live for thee'" (16). This song emphasizes Ida's transformation and the return to lived time: just as women should love children, nature should love humankind.[27]

Following the battle, Ida expands her concern for this child to a much broader concern for the many wounded. She commands her women to employ their gentle natures:

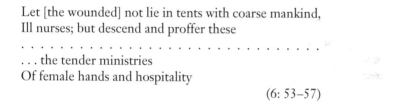

> Let [the wounded] not lie in tents with coarse mankind,
> Ill nurses; but descend and proffer these
> ·
> . . . the tender ministries
> Of female hands and hospitality
>
>   (6: 53–57)

Ida and her women renounce their studies for the sickroom. In *The Spectacle of Intimacy*, Karen Chase and Michael Levenson recognize Ida's decision to turn nurse rather than professor as the key moment in the poem's central reassurance: they write, "No one can doubt that the narrative offered its respectable readers the reassurance of a revolution not merely defeated but self-defeating" (123). Chase and Levenson explain the thematic import of the turn to nursing:

> Tennyson's Princess Ida arrives at her moment of domestic revelation when she descends to the battlefield, approaches the cross-dressing prince now lying badly wounded, and immediately decides to nurse his broken body. From this point, the resolution of the poem is assured. In the work of Sarah Ellis nursing also assumes a privilege, because the opportunity to tend the sick husband or the sick children gave inspiring grandeur to the woman-wife-daughter.[28] (131)

Of course, this is not the grandeur Ida had imagined for herself, nor the "grand, epic, homicidal" nature envisioned by the women of the frame. In nursing the wounded, Ida and her women exhibit care in place of cruelty, and they set aside their concern for generations to come in order to tend the individual lives of men. Ida nurses the prince, who hovers between life and death.

The indifferent nature of geological time moves in such grand strides that the prince would live and die in the shadow of a fleeting moment, yet Ida finally unleashes her own nurturing nature and devotes herself to saving the prince. She seems to relinquish her feminist goals entirely when she falls in love with the prince and dissolves her university.

As they did for Psyche, fond memories of a domestic past serve as the catalyst for Ida's final transformation. Ida reads the small relics of the prince's life and his love for her—her picture and a lock of her hair—and these inspire her to remember her affection for him:

> . . . she saw them, and a day
> Rose from the distance of her memory,
> When the good Queen, her mother, shore the tress
> With kisses, ere the days of Lady Blanche:
> And then once more she looked at my pale face:
> Till understanding all the foolish work
> Of Fancy, and the bitter close of all,
> Her iron will was broken in her mind;
> Her noble heart was molten in her breast;
>   (6: 95–103)

Traces from a personal past supplant the objects of deep time; the lock of hair replaces the sandy footprint as a signifier of Ida's aspirations. This collection of domestic *objets* persuades Ida to tame her wild nature. She sends each of her women "to her proper hearth" (7:2), and she thus asserts the traditional domestic narrative.

The prince describes Ida's acceptance of the domestic and denies any validity her university may have had when he describes the strong-willed Ida as her "falser self":

> . . . and all
> Her falser self slipt from her like a robe
> . . . .
> . . . when she came
> from barren deeps to conquer all with love
>   (7: 145–49)

As Venus emerges from the sea, the prince's Ida rises from the water, water which signifies the possibility of great change wrought through uniform processes over vast expanses of time, to embrace instead the possibility of familial love. Leaving geology behind when she rises from the "deeps," Ida also abandons her choice to remain "barren." Her rejection of "the first bones of time" and the cruel nature they represent is echoed in one of the poems Ida reads to the comatose prince. She reads of love found in a protective valley, not in the sublime landscape of geological time: "Nor wilt thou snare him in the white ravine, / Nor find him dropt upon the firths of ice / . . . / find him in the valley" (7:190–91; 195). She privileges gentle, protective nature over the violence implicit in ravines and ice. Tennyson's poem strips nature of her red teeth and claws and renders her instead a wife and mother who cares deeply about individual men.

Yet, the poem's comic resolution is self-conscious and holds within it the lingering seed of doubt that troubled from the start. In the end, Ida marries the prince but maintains her faith in uniformitarianism; his conflation of shadow and substance makes him her perfect mate. They present a pair of equals who will work together in small ways to affect change for women:

> And so these twain, upon the skirts of Time,
> Sit side by side, full-summed in all their powers,
> Dispensing harvest, sowing the To-be,
> Self-reverent each and reverencing each,
> Distinct in individualities,
> But like to each other even as those who love.
> Then comes the statelier Eden back to men:
> Then reign the world's great bridals, chaste and calm:
> Then springs the crowning race of humankind.
> May these things be!
>
> (7: 271–79)

These lines describe a union that represents a compromise between Ida's feminist ideals and the convention she reluctantly assumes. The narrating poet concludes the frame story with the assertion that the tale traces a "strange diagonal" between comic and tragic, between traditional and revolutionary. Indeed, the marriage seems to balance on this diagonal.

The story's denouement also rests in the liminal space between cruel nature and nature domesticated. The description of the "statelier Eden" suggests a happy ending, a suppression of nature's indifference and a return to a reading of nature governed by the Bible rather than geology. The evocation of a prelapsarian nature reflects the Victorian fall from the innocence that prevailed before knowledge of geological time. Yet Eden "comes" in the future, Ida and her husband sow the "To-be," and the finale ends with the cheerful, "May these things be!" These things are not yet. The speaker asserts that Eden will return, but his assertion is only a hope—the hope that drives the whole poem. Just as Ida's "wild nature" has been tamed, domesticity may also tame nature's wildness, but the poem cannot contain such taming as a fait accompli. The domestication of nature is finally nothing more than a dream, a fairy tale, an almost desperate attempt to "shape the shadow Time." Even Ida questions her conversion. Her last words in the poem raise doubts about her transformation: "I seem / A mockery to my own self. Never, Prince, / You cannot love me" (7:316–18). She challenges the possibility of love between them, and she thus challenges also the possibility of the familial narrative. Finally, the irrepressible light of geology dissolves the shadows so desperately constructed. No narrator can totally revise the frightening truth that time is a construct and nature is cold.

Concluding with Lilia

The poem concludes with a return to the frame narrative and to the festival on the grounds of the estate, where the medley has expanded to include France: the storytellers gaze "far beyond, / [at,] Imagined more than seen, the skirts of France" (47–48). These lines were added to the 1850 edition, written after the 1848 revolutionary troubles in France. At this time, the fear of revolution in England remained very real, and the threat of a social revolution took on greater proportion in such a context. Indeed, the speaker looks specifically to the "skirts" of France, suggesting a female revolution originating across the channel but irrevocably altering British life in its wake. A Tory member of the party proceeds to describe an orderly and reasonable process of social change that is disrupted by social unrest:

> God bless the narrow sea which keeps her off,
> And keeps our Britain, whole within herself,
> A nation yet, the rulers and the ruled—
> Some sense of duty, something of a faith,
> Some reverence for the laws ourselves have made,
> Some patient force to change them when we will,
> Some civic manhood firm against the crowd—
> But yonder, whiff! There comes a sudden heat . . .
> (Conclusion: 51–58)

The "sudden heat" suggests the catastrophic possibilities associated with the Huttonian theory of the Earth's fiery core. The Tory connects social revolution to geological catastrophism. He expresses a genuine fear of "mock heroics stranger than our own" and the "revolts" and "revolutions" they may inspire (64–65).

The poet assuages fears of revolution, whether political, social, or scientific, with an expression of faith in science, particularly uniformitarianism. Change happens so slowly that it is not disruptive, and, though one may be forced to doubt the biblical narrative, one can have absolute faith in the constancy of the universe. Steady work over time reforms the surface of the Earth, and steady work over time reforms humanity's social constructions of reality. The poet suggests, ". . . maybe wildest dreams / Are but the needful preludes of the truth" (73–74). Such dreams need not threaten the social order for they are part of an order so much larger and so much more powerful that their significance cannot be overestimated. The troublesome rumblings of F. D. Maurice and other advocates for women's education do not rend the Earth; they are only minuscule vibrations that will slowly, gently reshape society over millions of years. Such gradualism appeals to Tennyson and his contemporaries. The speaker finally accords a sacred value to the uniformity displayed at the festival: "The sport half-science fill[s] me with a faith" (76). This faith governs Ida's reform project, and it offers assurance at the poem's end. Of course, it is the frame-poet who finally looks to the conjunction of past, in the items on display at the house museum, and present, in the science on display at the festival, and crafts from these temporal fragments, as well as from the fragments of story, a narrative that as a medley offers a coeval vision of the age. Moreover, in assembling these fragments he not only constructs a model of his times (and of Time), he also offers a palliative to the anxiety those very fragments inspire.

In Lilia, the frame's young feminist, we find the long result of time and Ida's work. She responds to the feminist position, but she does so with caution. She looks to her knowledgeable Aunt Elizabeth to understand the woman question: "'You—tell us what we are'" (34). She hopes that "theories out of books" (35) will provide some clarity, but finally it is the festival with its scientific medley that offers Lilia a model of reform. The internal narrative does not inspire Lilia to reject home and found a women's university; rather, the poem leaves Lilia confused, and her confusion reflects the gradual change Ida and uniformitarianism teach us to expect. Lilia's final gesture is to lift the woman's costume from the statue of Sir Ralph: she removes the guise to discover the substance beneath. For Lilia and her Victorian contemporaries, uncovering substance is an ongoing process. They may not yet see the reality of deep time unobscured, but they have begun the long process of excavation. Women in 1850 may not be ready for the independence Ida sought, yet neither do they celebrate Ida's return to the home. Finally, it seems that Ida's uniformitarian project succeeds. She plants her foot in time, and "young Lilia" is the green bud that grows from Ida's seed. This fairy tale does have a happy ending. Nature may be wild, but she is not indifferent. She takes generations to cultivate a species, but she does cultivate it. *The Princess* strives to offer a balm to the existential angst provoked by geological time: in its broadest gestures, it fails, but in the smaller claims of a teardrop, it assuages the concerns of a generation whose temporal narrative has been shattered. The poem concludes with Lilia and all women, even all humankind, "rising quietly" (116).

4

Accidental Archaeology
in London and Pompeii

These fragments I have shored against my ruin . . .
—T. S. Eliot, *The Waste Land*

Geology brought the trace—both the product and the sign of deep time—to the foreground of nineteenth-century experience, and the trace became the foundation of studied reconstructions of the past. Wielding a fossil and a pen, the geologist asserted his own narrative authority despite the apparent indifference of nature and time to human experience. Both alarmed and invigorated by geological discoveries, Tennyson, and many others, struggled to tell stories in which individual people find a way to matter in deep time. For all writers influenced by geology, the trace oriented the quest, whether a quest for knowledge or the far more important quest for significance. Archaeology offered an attractive parity between the trace and individual experience. Rather than suggesting nature's total indifference to humanity, the objects excavated by archaeologists revealed the extraordinary power of certain items to endure. These traces offered a glimpse of a past civilization, even a past life. Monuments may decay, but they continue to pronounce bygone glories; even personal trifles preserved over time recall the small sentiments of individuals dead and forgotten centuries ago. Reduced to traces, men and women must necessarily be absent, yet the mark of their absence is a form of preservation. At the site of the trace—whether artifact, remains, or ruins—archaeology explores the rocky meeting place of mortality and immortality; such places allow the observer to face and possibly conquer time. This chapter examines the emphasis on the individual and his or her struggle against time's destruction at two archaeological sites: Pompeii and London.

Both Pompeii and London preserve remnants of the Roman Empire, yet are distinct from more monumental sites of Roman ruin: they are familiar, even quotidian. In contrast, most of the archaeological work undertaken in the nineteenth century involved the excavation and co-option of monuments, religious structures, and the material associated with kings, pharaohs, and other exceptional individuals. The laundry lists at Pompeii and the pottery shards found in London hardly deserve to share the label artifact with grand objets d'art like the treasures found in pyramids or the Elgin marbles. Widely considered the most valuable artifacts collected and displayed in the nineteenth century, the Elgin marbles epitomize the emphasis on the monumental: as Ian Jenkins contends in *Archaeologists and Aesthetes in the Sculpture Galleries of the British Museum 1800–1939*, "The acquisition of the Elgin marbles is arguably the single most important event in the history of the British Museum and certainly in the course of nineteenth-century European *Antikenrezeption*" (19). These marbles from the Parthenon were of extraordinary beauty, and they became the standard against which other archaeological finds were judged.[1] In contrast, Pompeii and Londinium are not known for their monuments or their aesthetic value, but rather for their ticket stubs, hair combs, and sandals, all traces of individual experiences, objects of little value or importance in their own time and notable in the nineteenth century only because of their striking familiarity.[2] Faced with the familiar and the quotidian, the nineteenth-century visitors to these sites shared the experience of Henry Knight, clinging to a cliff, face-to-face with a trilobite: time closed up like a fan before them. Past and present slammed into one another, and the individual, though hanging on for dear life, was suspended outside time. In this conflation of past and present, we see an example of what Hegel called reflective history:[3] rather than considering the past as passed, the Victorians emphasized the interrelation between past and present, an interrelation dependent on the quotidian traces so characteristic of each site.

Pompeii and Londinium both function as bridges, connecting past and present through recognizable objects, particularly at Pompeii, and through shared space in London. The physical proximity of Londinium, right in the middle of nineteenth-century London's thriving commercial center, thrust its remains into the daily experience of the public. The traces of Londinium uncovered in nineteenth-century London were found, quite literally, in the backyard. Well-known locations throughout the city, such as the Royal Exchange or the Strand more generally, revealed traces of the Roman past. Moreover, most of the archaeology in London was accidental: the discovery of objects was a by-product of excavation and construction intended to improve the urban landscape. Thus, the remains of Londinium were intimately associated with the investment in London's future. Much of the writing about Pompeii also focuses on the future of London, as writers, ranging from the earnest to the light-hearted, imagine what would become of London were it suddenly destroyed (or preserved) by a volcanic eruption or some other devastation. Though Pompeii was geographically distant from the British writers and travelers I consider here, it was culturally proximate. That is, the signs of everyday life struck nineteenth-century observers as very much like their own: children's games were similar, as was a lady's dressing table, and even graffiti rendered the city a phantasm of future London. Of course, one can study the archaeological sites of Pompeii and Londinium without emphasizing the familiar; one can draw maps and write histories and give very little thought to the lives lived in those places. Yet, I contend that it was the accessibility of individual experience that made these particular sites so appealing in the nineteenth century.

In addition to their emphasis on the familiar, Pompeii and Londinium offered visitors an opportunity to reflect on the likeness of the Roman Empire to their own expanding empire in the nineteenth century. In *The Victorians and Ancient Rome*, Norman Vance explains, ". . . ancient Rome, variously constructed or reconstructed, could be not just part of the dead past but a vision and an idea transcending its original context: in politics and in the private sphere, in life and literature, Rome presented challenging paradigms and reference-points, ways of making sense of a chaotic and volatile present, and possible models for the undisclosed future" (5).[4] Vance offers Rome as a tool used by the Victorians for engaging with and comprehending their own time, and this is the approach to Rome that most interests me. Important to this consideration is the notion that Rome was the lesser of the classical cultures: in his intellectual history of the Victorian period, Robin Gilmour explains that admiration of Rome was typical of the eighteenth century, whereas the nineteenth century was dominated by Hellenism. Jenkins confirms this view in his discussion of the efforts in the nineteenth century to organize the sculptures in the British Museum in a progressive, chronological order. Jenkins explains: "In the traditional eighteenth- and nineteenth-century hierarchy of ancient art, Egypt represents the primitive forerunner of Greece, while Rome is seen as the manifestation of an inevitable decline" (102). It is perhaps because of the empire's decline, its fall from greatness, that Rome resonated with the Victorians, poised at a moment in history where they were keenly aware of the possibility of their own fall.

Though I touch on the compelling nexus of archaeology, its imperial context, and national identity in the following discussion, my primary emphasis, as in chapter 2, is on the individual. In *Voices in the Past: English Literature and Archaeology*, John Hines emphasizes the role of the individual in the creation of archaeological sites; he explains how the material culture of interest to archaeology is a product of individual decisions—where to put the jewels, how to house the dead, and so on. In this way, archaeology is fundamentally distinct from geology, which studies the location of rocks and fossils positioned through natural processes. However, in my view, both sciences emphasize the interpretive power of the individual: recall the observer at the margins of Lyell's frontispiece. I agree with Hines that individual decisions are important in the construction of remains, and, further, I assert that the individual is also the central figure in interpreting remains. Referencing Foucault, Hines distinguishes an archaeological from an epistemological approach to history. The archaeological approach has to do with the unconscious construction of material history, and the epistemological approach concerns the conscious interpretation and construction of historical narrative. In this chapter, my focus is on the meeting point of the archaeological and the epistemological. The former derives from the daily actions of the individual, and the latter is produced by an individual interpreter or writer. In both cases, the individual is fundamental to the production of both artifact and narrative. Even the remains themselves seem generally less important than the narratives of pursuit and the stories crafted out of recovered fragments. The literature that comes out of archaeology thus connects the young discipline also with the new historiography that so emphasized narrative.

Throughout this chapter, "archaeology" describes a developing field, one which, unlike geology, remained ill-defined until late in the nineteenth century. Archaeology can be considered at once the scientific arm of antiquarianism and the branch of geology focused on human artifacts. In the nineteenth century, the tyro discipline encompassed activities as diverse as sketching a ruined abbey, plundering an Egyptian tomb, and dredging Roman bronzes from the Thames. Associated with both Romanticism and science, archaeology had a single defining feature: the pursuit of material remains of past civilizations. Antiquarianism has a long and rich history, dominated by the local and national societies whose meetings offered a forum for collectors, historians, and early archaeologists to tell their stories.[5] In contrast, archaeology did not really become a science akin to the one we know today until the middle of the nineteenth century when its connection to geology was fully realized; even then methods were not at all standardized. Glyn Daniel states the case very clearly: "There could be no real archaeology before geology" (24). The work that falls under the heading antiquarianism was more or less scientific, depending on the individuals and the subjects involved. Digging into a barrow required at least rudimentary methods of excavation, whereas a visit to a ruined abbey was often more Romantic and involved sketching and descriptive prose, rather than anything resembling scientific examination.[6]

The figure most historians of archaeology identify as the best of the proto-archaeologists is William Stukeley (1687–1765), and a brief look at his work shows archaeology's antiquarian origins and its movement toward science. Stukeley worked throughout Britain on Celtic and Roman sites and is best known for his work on Avebury and Stonehenge,[7] though his reputation is complicated by an obsession with druids that characterized his later writings about Celtic remains and has made it difficult to consider his contributions as scien-

tific.[8] Nonetheless, most historians single out Stukeley as advancing archaeology significantly. Glyn Daniel writes in *A Hundred and Fifty Years of Archaeology*, "The finest example of the Romantic British archaeologist was, of course, William Stukeley" (23), and in *The Early Barrow-Diggers*, Barry Marsden claims, "Stukeley is regarded as the first great English field archaeologist and antiquarian pioneer, and his acute observation and recording contributed much to the growing interest in archaeological field studies . . ." (3). Considering what set Stukeley apart from his contemporaries, Van Riper explains, "Stukeley firmly believed that the study of Britain's antiquities should be practiced in the field" (19). This emphasis partnered with the rapidly increasing accessibility of the field, through railways most of all, made archaeology into a discipline not quite a science but ready to become one. As a digger, Stukeley was no more scientific than his contemporaries, but as a recorder of information, one who considered the whole of a site and all its artifacts in situ, Stukeley brought discipline to mid-eighteenth-century archaeology and began to shift the field from its antiquarian origins closer toward its geological ends. Still, stratigraphy and classification of artifacts were not within Stukeley's purview.

If Stukeley represents the beginning of archaeology as a discipline, Augustus Henry Lane-Fox Pitt-Rivers (1827–1900) represents its complete transformation into a science. Pitt-Rivers relied on scientific classification systems to understand his discoveries, and, compared to his predecessors, he was meticulous in his excavations. In his biography of Pitt-Rivers, Mark Bowden explains that Pitt-Rivers had no professional methodology to adopt: "Fieldwork in British archaeology had developed unsystematically through the work of a number of more or less gifted individuals from . . . William Stukeley onwards. There was therefore no school of fieldwork into which the General's career can be fitted" (57). Bowden goes on to explain that Pitt-Rivers, despite his having to cast about for a methodology, considered himself a scientist and "did not associate with historians" (160). His scientific inclinations mark Pitt-Rivers as far more an archaeologist than an antiquary, a distinction impossible to make earlier in the century. Indeed, with Pitt-Rivers, we see the final move away from antiquarianism and its associations with the Romantic view of the landscape. Finally, both geology and archaeology relied on the meticulous excavation of material remains, the careful notation of their condition and situation, the delicate removal of the artifacts, and perhaps even some strategic reconstruction. Moreover, both excavatory sciences depended on the interpretive authority of the scientist to make sense of his finds and to write the past.

The connection between geology and archaeology was most obvious at prehistoric sites that contained remains of both humans and extinct animals. The catastrophist approach to geology led many who encountered such evidence to conclude that these occurrences were accidental: for example, Romans having imported elephants was an easy explanation for the coexistence of human and mammoth bones.[9] However, geology eventually began to interpret this evidence correctly. Daniel and many others identify the excavations by Falconer and Pengelly in England and by Boucher de Perthes in France as the key events that "persuaded the scientific world to accept the contemporaneity of man and extinct animals" (Daniel 57). When Charles Lyell addressed the Geological Section of the British Association meeting at Aberdeen in 1859 and announced that he was "fully prepared to corroborate the conclusions . . . recently laid before the Royal Society" (qtd. in Daniel and Renfrew 37), the paradigm definitively shifted. Speculation could then turn to how old humans were, rather than whether or not they were, in fact, old.[10] Once this conclusion had been reached and was accepted by the scientific community, it entered archaeology as a given, and the discourse of that discipline shifted to meet the new understanding that human prehistory was a subject for excavation. Historians of archaeology like Daniel and Van Riper consider the development of prehistory central to the movement of archaeology toward science and away from antiquarianism: it made evident the need for geological methods in the study of human remains, and once it was clear that those methods should be applied to the excavation of prehistoric sites, it was only logical to recognize the value of employing similar methods at historic sites.[11]

Geologists were the first to note the connection of the two fields. In 1851, Gideon Mantell described the relationship between archaeology and geology in an essay that assessed the evidence of prehistoric human remains. Mantell sees the two sciences as fundamentally the same, differing only in the antiquity of the objects of study:

> . . . as the antiquary, from a fragment of pottery, or a mutilated statue, or a defaced coin,—objects intrinsically valueless, but hallowed by the lapse of ages,—is enabled to determine the degree of civilization attained by a people whose origin and early history are lost in remote antiquity; so the geologist, from the examination of a pebble, or a bone, or a shell, may ascertain the condition of our planet, and the nature of its inhabitants, in periods long antecedent to all human history or tradition. (235)

Mantell articulates clearly how the methods of archaeology and geology were so closely related; he is also quite eloquent on the subject of traces, "objects intrinsically valueless, but hallowed by the lapse of ages" and rife with signification. Writing just a few years earlier, Hugh Miller (1802–1856),[12] also a geologist, reported on his visit to the town museum at Newcastle-on-Tyne in his *First Impressions of England and Its People* (1847) and stressed the relationship of geological and archaeological artifacts:

> . . . as I passed, in the geologic department, from the older Silurian to the newer Teritiary, and then on from the newer Tertiary to the votive tablets, sacrificial altars and sepulchral memorials of the Anglo-Roman gallery, I could not help regarding them as all belonging to

one department. The antiquities piece on in natural sequence to the geology; and it seems but rational to indulge in the same sort of reasonings regarding them. They are fossils of an extinct order of things, newer than the Tertiary—of an extinct race—of an extinct religion—of a state of society and a class of enterprises which the world saw once but will never see again. (qtd. in Torrens, 35–36)

Like the museum Miller describes, the seamless progression from geological stone artifacts up to the decorative items of civilized societies was displayed in the archaeological collection of Charles Roach Smith, a noted London archaeologist. (See figure 4.1.) Though Mantell and Miller both articulate the connection between archaeology and geology, they do so with foresight, as it would be another ten years before the connection was generally made.

The extraordinary preservation of prehistoric humans offered a paradox: such remains forcibly reminded observers about the fragility of human life, but they also inspired imaginings about the potential of all humans to endure the cruel ravages of nature and time through conversion into immortal artifacts. Remains, whatever else they may be, do remain. Writing in *The Wonders of Geology*, Mantell makes the vital connection between the human remains discovered by geologists and the potential preservation of human cultures of the present:

The occurrence of human skeletons in modern limestone . . . incontestably prove[s] that enduring memorials of the present state of animated nature will be transmitted to future ages. When the beds of the existing seas shall be elevated above the waters, and covered with woods and forests—when the deltas of our rivers shall be converted into fertile tracts, and become the sites of towns and cities—we cannot doubt that in the materials extracted for their edifices, the then existing races of mankind will discover indelible records of the physical history of our times, long after all traces of those stupendous works, upon which we vainly attempt to confer immortality, shall have disappeared. (124)

Mantell may have recalled Percy Shelley's "Ozymandias" (1818). Shelley famously captures the irony of the ruined monument in "Ozymandias," where the "lone and level sands stretch[ing] far away" belie the arrogant words of the long-forgotten king: "Look on my works, ye mighty, and despair" (14, 11).[13] The reader does despair when he discovers the fragility and insignificance of great rulers and great empires: if kings disappear, what of regular men and women? Archaeology's excavation of destroyed cities and ruined cultures made undeniably apparent the smallness of individual lives. Edward Bulwer-Lytton reminds his readers of their insignificance in *The Last Days of Pompeii* (1834) when Arabeces describes the irrelevance of human life in a uniformitarian model of change: "Of that which created the world, we know, we can know, nothing, save these

Fig. 4.1. "The Museum of C. Roach Smith, Esq.," courtesy Guildhall Library, City of London.

attributes,—power and unvarying regularity; stern, crushing, relentless regularity, heeding no individual cases, rolling, sweeping, burning on, no matter what scattered hearts, severed from the general mass, fall ground and scorched beneath its wheels" (62). The expanded time scale teaches the painful lesson that no person and no civilization can endure throughout eons and eons of incremental but monumental change. Yet, Mantell does describe individuals—and cultures— that endure. He and Shelley both emphasize the failure of stupendous works to stand the test of time, but Mantell sees hope in a different sort of ruin. The humans who crafted the works will remain, fossilized, for ages to come, and they will confer immortality upon themselves even as their monuments fail them.

While monuments crumble into oblivion, their power to mean something endures as archaeologists and others examine and interpret traces. Fabian's coeval cultural analysis, which I extend to a coeval temporal analysis, has a corollary in uniformitarianism. In *A History of Archaeological Thought*, Bruce G. Trigger explains:

> Archaeology is a social science in the sense that it tries to explain what has happened to specific groups of human beings in the past and to generalize about processes of cultural change. Yet, unlike ethnologists, geographers, sociologists, political scientists, and economists, archaeologists cannot observe the behavior of the people they are studying and, unlike historians, most of them do not have direct access to the thoughts of these people as they are recorded in written texts. Instead archaeologists must infer human behaviour and ideas from the material remains of what human beings have made and used and of their physical impact on the environment. The interpretation of archaeological data depends upon an understanding of how human beings behave at the present time and particularly of how this behaviour is reflected in material culture. (19)

Like geology, archaeology relies on uniformitarianism as an interpretive tool. The Victorians understand past human behavior through their own very particular present-day lens, and thus they craft a reflective history heavily infused with nineteenth-century sensibility. They find past and present coexistent in an anachronistic space, a space that the interpreting individual has the power to control and to shape into narrative form.[14]

Pompeii—The City of the Dead

Unlike many archaeological finds, the ruins at Pompeii were not intentionally preserved; rather, the city and all its private spaces were suddenly petrified with very little warning. In his description of archaeology, Hines emphasizes the deliberate acts of individuals who determine to preserve their valuables in particular ways: Pompeii illustrates the fallacy of such determination, and reveals instead the individual decisions of everyday life. Because it was unplanned, the preservation of Pompeii offers rare insight into the daily lives of the Romans, and the remains of private life especially intrigued nineteenth-century observers who so highly revered the private sphere in their own culture. The ruin of the walls holding public and private apart, segregating the classes, and separating men and women, titillated nineteenth-century readers and invited them to imagine the destruction of their own cities and social values. Enough remains at Pompeii to offer considerable insight into Roman life, yet the remains are fragmented and require interpretation. Thus, extending Hines's emphasis on the individual, individual acts of discovery and reconstruction are central at Pompeii.

Excavation of Pompeii began in 1748 and proceeded sporadically throughout the eighteenth and nineteenth centuries. In *Gods, Graves, and Scholars,* C. W. Ceram explains how the work on Pompeii began as the fanciful whim of the King of the Sicilies and his wife. The superficial interest of the royal couple prevented systematic excavation, although important discoveries were made nonetheless. The early approach was to look hurriedly for anything of great interest or value; when a particular area failed to offer such objects quickly, the site was abandoned and digging commenced elsewhere. Eventually, the whole of Pompeii was deserted in favor of Herculaneum, which initially offered more impressive artifacts. The early history of Pompeii's excavation reveals that archaeology was then far more sport than science, and this approach continued to characterize excavation of the site well into the nineteenth century. In his history of the excavation of Pompeii, published in *The Quarterly Review* in 1864, Austen Henry Layard describes the practice of reburying interesting objects so that visitors could have the satisfaction of uncovering them. These faux excavations demonstrate the careless method employed at Pompeii and also the emphasis on impressive objects at the expense of those items that appeared less interesting. Furthermore, the staged excavations Layard describes highlight the importance of the dig and the discovery. The impressive objects had already been discovered, so the clear focus of these shows was the faux-archaeologist him- or herself.

The exhibitions of discovery staged at Pompeii seem clownish when compared with the serious archaeology that would develop later in the century, but if we probe these archaeological stunts more deeply, we discover that they represent a particular attitude toward the past. Moreover, they represent a particular attitude toward the archaeologist. In his discussion of Assyrian artifacts, Shawn Malley emphasizes the performance aspects of archaeology theorized by Mike Pearson and Michael Shanks. This theory stresses the central importance of the archaeologist as an interpreter. Malley summarizes: "Archaeology as performance is, then, a way of articulating the relationships we have with the past, and a mode of understanding the roles of artifacts in social practices and identities. In the face of loss, decay and forgetfulness, performance takes up the material past and reanimates it in and for the present" (2004, 3). Here again are echoes of Hegel's reflective history, a history as, if not more, interested in the present as in the past. The false excavations Layard describes can surely be considered performances: the scene is set for the excavator to assume center stage and, with the apparent ease of an expert, make a remarkable find. A true excavator has already unearthed a collection of artifacts, some interesting and others not, and has selected certain items that visitors will find impressive. The visitors then bypass the struggle of locating a promising site and the labor of digging for an extended time. The story these performances tell about archaeology is that the past is at the disposal of the present. It lies near the surface, its valuable and interesting objects reflecting positively on those who discover them. Of course, the performance has another, more interesting layer: the individuals who lay the ground, literally, for the discovery are themselves staging the relationship with the past. They emphasize the value of single objects over a whole scene or environment, and, though they surely appreciate the real work of excavation and the holistic view of a place like Pompeii, they are complicit in valuing the excavator above the site.

Though excavation at Pompeii may have been somewhat more advanced than elsewhere, archaeological excavations in the first half of the nineteenth century were generally conducted in a haphazard manner that emphasized singular discoveries.[15] In his study of barrow digging, Marsden describes the imprecise nature of the excavations in Britain. He writes:

> Many of the barrow-openers were great collectors, possessing strong acquisitive instincts which led them to amass great assemblages of relics. Many hundreds of them worked in a careless fashion, leaving no account of their labours, and thus denying us the information resulting from the pillaging of innumerable mounds. Many did too much too quickly—again faults of the era in which they lived. (vii)[16]

Daniel dramatically describes the damaging methods of Giovanni Battista Belzoni (1778–1823), who famously brought back to England some amazing Egyptian finds:

> He broke into one tomb, crashing his way through antiquities; "every step I took I crushed a mummy in some part or other," he declares. "When my weight bore on the body of an Egyptian it crushed like a band-box. I sank altogether among the broken mummies with a crash of bones, rags and wooden cases. . . . I could not avoid being covered with bones, legs, arms and heads rolling from above." (155–56)

Belzoni's account places the archaeologist—himself—center stage, at once crushing the artifacts and almost becoming one of them. Belzoni was one of the most renowned archaeologists of the early nineteenth century, surpassed in the next generation by Layard. Layard conducted fruitful excavations in Nineveh and wrote what is considered to be the first archaeological best seller of the nineteenth century, but, as Daniel explains:

> . . . Layard himself did little for the preservation of his own finds. *Nineveh and Its Remains* abounds in the phrase reminiscent of the earlier barrow diggers in England, and the early nineteenth-century excavators of Egyptian mummies: "entire when first exposed to view, it crumbled into dust as soon as touched." Thus copper helmets, iron armour, copper vessels, painted frescoes, ivories, all "fell to pieces almost immediately on exposure to the air." (155)[17]

In his excellent history of archaeology, Kenneth Hudson does not even consider the early part of the nineteenth century, presumably finding nothing there that fits his definition of archaeology.

Yet, studies of Pompeii and visits to the site were popular throughout the nineteenth century, and these generally involve marveling at what has been preserved rather than what has been lost. Pompeii is an especially rich subject for a close examination of the archaeology of this period for there was a sense that anyone could contribute to the interpretation of this particular archaeological trace, and these contributions so routinely celebrate the individual and the quotidian. With his Romantic endorsement, Sir Walter Scott secured Pompeii's popular reputation as a sentimental memorial to private life: upon first seeing the ruin, he exclaimed, "The City of the Dead," emphasizing the site's powerful amalgam of the morbid, the sublime, and, perhaps most appealing, the personal. Popular interest in Pompeii always focused on the people killed by the volcanic blast and the relics that offer insight into their private, final moments. Nineteenth-century writers reconstructed Pompeii's fragments into a story of personal struggle, pathos, and tragedy. While, in fact, Vesuvius spewed smoke and ash for days before its spectacular eruption, giving most people a chance to flee Pompeii in good time,[18] the narrative related throughout the nineteenth century features a sudden, surprise explosion that caught the citizens of Pompeii unawares and petrified them forever in their tracks. Many writers enhance this narrative by taking the liberty of describing the scene at Pompeii as the original excavators would have found it, not as contemporary visitors would actually see it. By the early nineteenth century, any loose objects—pottery, foodstuffs, jewelry, and so on—had been removed to the Museo Burbonico in Naples. Nonetheless, most descriptions of the scene romanticize not only the last day of Pompeii but also the contemporary ruin; popular writers suggest that the visitor will see homes just as fleeing Pompeians left them, with dinner on the table and daily debris on the floor.

In *Re-Constructing Archaeology*, Michael Shanks and Christopher Tilley stress the importance of storytelling, of recorded observations. They write,

"Archaeologists observe the traces of the past then record and write about them. Archaeologists produce texts. Archaeology depends on texts. . . . Its function is archival" (16). This description of archaeology offers a compelling corollary to the distinction between studies of literate and preliterate societies. For those societies that produced their own texts, some consider archaeology a mere complement, confirming or elaborating whatever history was recorded. However, Shanks and Tilley emphasize the texts produced by the archaeologists rather than any extant texts from the culture under examination. To illustrate the distinction, I offer the letters of Pliny the Younger, the only surviving eyewitness accounts of the eruption of Vesuvius. In the nineteenth century, Pliny's descriptions were known widely and referenced in many treatments of Pompeii; however, Pliny's voice is almost always subordinated to the textual authority of the nineteenth-century writer. Pliny wrote: ". . . the cloud was rising . . . its general appearance can best be expressed as being like an umbrella pine, for it rose to a great height on a sort of trunk and then split off into branches . . ." (Radice 166). Paraphrasing and with no acknowledgment of Pliny's letter, Layard wrote in the *Quarterly Review*: "Suddenly, and without any previous warning, a vast column of black smoke burst from the overhanging mountain. Rising to a prodigious height in the cloudless summer sky, it then gradually spread itself out like the head of some mighty Italian pine" (313).[19] The original text is suppressed, used as a source by nineteenth-century writers who co-opt even its simile but represent the description as their own, asserting their own voice over the single lasting voice from the past. William Makepeace Thackeray (1811–1863) offers a more explicit example of Pliny's inferior status in *The Newcomes* (1853): "The young man had read Sir Bulwer Lytton's delightful story, which has become the history of Pompeii, before they came thither, and Pliny's description, *apud* the 'Guide Book'" (418). Thackeray presents *The Last Days of Pompeii*, a nineteenth-century novel, as the definitive history of Pompeii, and though he mentions Pliny's letter, he subordinates it to Bulwer-Lytton's text and even further diminishes its authority by couching it within a travel guide.[20] Nineteenth-century writers and archaeologists were less interested in official histories of Pompeii, which Pliny aimed to provide, than in their own interpretations that emphasized the quotidian. Shanks and Tilley would argue that such interpretations are the substance of archaeology.

The repeated emphasis on a Pompeii exactly preserved just as it had been abandoned adhered to a philosophy of history that accorded the historian or the archaeologist the authority to bring the past to life. Barthold Niebuhr (1776–1831), a highly influential Danish historian whose *History of Rome* (1811–1832) expanded the historical discipline, legitimized the art of reanimating the past. Describing Niebuhr's importance for Victorian historiography, Linda Dowling writes, "Niebuhr . . . believed . . . that philology collapsed the distance between ancient and modern times by opening the very minds of vanished generations to modern understanding. The modern historian of

Rome could greet the ancient Romans as contemporaries" (584–85). Dowling goes on to quote Niebuhr's endorsement of reanimating the past, a passage also cited by many Victorians, notably by Charles Lyell in *Principles of Geology*:

> He who calls departed ages into being enjoys a bliss like that of cre-
> ating: it were a great thing, if I could scatter the mist that lies upon
> this most excellent portion of ancient story, and could spread a clear
> light over it, so that the Romans shall stand before the eyes of my
> readers, distinct, intelligible, familiar as contemporaries, with their
> institutions and the vicissitudes of their destiny, living and moving.
> (qtd. by Dowling 585)

Van Riper challenges the importance of Niebuhr for archaeology, noting that reconstructing the past was common practice in geology. He mentions the drawings that attempted to illustrate prehistoric life and places like the di-nosaur park at the Crystal Palace in Sydenham, which offer even more explicit reconstructions.[21] He writes: "The goal of creating such reconstructions was rooted not in history but in natural science. In the early 1830s, a group of En-glish geologists led by William Buckland added a new dimension to the study of Earth history by pioneering the reconstruction of ancient environments" (31). Archaeology certainly followed geology in its proclivity for reconstruc-tion, but given the extent of these reconstructions in the early years of the nineteenth century and archaeology's great debt to antiquarianism and history during those years, it seems fair to credit Niebuhr along with the geologists for the effort to reanimate the past. Of course, the stress Niebuhr placed on reanimation corresponds interestingly with Hegel's view of history: if one strives to experience the past exactly, then this seems to correspond to Hegel's original history; however, if one thus brings the fully realized past into the present, then one is certainly reflecting on the parity between past and pres-ent. As the temporal gap closes, perhaps Niebuhr and his like-minded con-temporaries crept toward Hegel's philosophical history. Ironically, the study of time might lead to its erasure.

Edward Bulwer-Lytton (1803–1873), Thomas Babington Macaulay (1800–1859), Charles Dickens (1812–1870), and scores of others writing in Niebuhr's wake attempted to bring Pompeii to life through their histories, fic-tional or otherwise. Bulwer-Lytton describes his aim in the introduction to *The Last Days of Pompeii*: ". . . it was not unnatural, perhaps, that a writer . . . should feel a keen desire to people once more those deserted streets, to repair those graceful ruins, to reanimate the bones which were yet spared to his survey, to traverse the gulf of eighteen centuries, and to wake to a second existence— the City of the Dead!" (xiii). Similarly, Stephen Middleton concludes his 1835 poem with the claim that he has brought the city to life, "The Poet . . . / Renews with life each long deserted scene,"[22] and Thomas Gray (1803–1849), the author of the 1830 novel *The Vestal*, asserts his power to reanimate the dead city, "On entering the limits of the sepulchre of a city, we seem to have stepped back over a space of two thousand years. . . . The curtain that separates the past from the present is taken away, and we breathe and move among realized dreams and fables" (n.p.). Through attention to vivid, quotidian detail—detail that echoed the daily clutter of Victorian lives—many borrowed Niebuhr's historiography to make Pompeii live again. Those who write on Pompeii assume the authority not only to report the city's history but also to bring the dead back to life and to blur the distinction between past and present.

Many writers allude to Sleeping Beauty to describe their magical powers to reverse the volcano's eruption and the passage of time. Like the castle in this fairy tale and also strikingly like McClintock's anachronistic space, the city of Pompeii seems frozen, its inhabitants sleeping yet quivering still with life; it falls to writers to awaken them. The introduction to *The Vestal*, for example, describes the reanimation of Pompeii in these terms:

> There is a sensation of inexpressible interest and delight in the feeling with which we view this astonishing preservation of antiquity; almost realizing the fairy tale that delighted our childhood, in which the sleeping princess . . . with her whole court was put to sleep; and when, after the lapse of many years, the enchantment was broken, they all awoke as young as when they first slept, and all instantly recommenced the avocations that occupied them at the moment of the enchantment, unconscious of the lapse of years, or of the changes in the world around them. (n.p.)

This author describes being petrified for two thousand years as an "enchantment," and presumes to reanimate the dead with his prose. Similarly invoking Sleeping Beauty, "Behind the Scenes at Vesuvius," published in *All the Year Round* in 1863, characterizes Pompeii's buildings as "all struck to silence like the Sleeping Beauty, only for a great many hundred years instead of one; and, in our day, so restored to light and life, that we see what the townspeople were doing in the house and in the street, in the month of August AD 79" (473). As the geologist assumes the authority to write time, the archaeologist, in this fairy tale reading of the discipline, similarly authors the past.

Contributing to the paradoxical notion that Pompeii was both the "City of the Dead" and a city of the living was the peculiar fact that nineteenth-century Pompeii did have inhabitants. In "A Day at Pompeii" (1855), the author mentions that the king is "doing something in the way of repeopling the city" by authorizing selected soldiers to choose their residences among the ruined houses in the city (727). Commenting on the same in "The City of Sudden Death" published in *Household Words* in 1852, John Delaware Lewis (1828–1884) notes the oddity of seeing laundry hung up to dry amongst the remains of the dead city (173).[23] While nineteenth-century visitors to Pompeii acknowledge the contemporary inhabitants of the ruined city, they take no substantive interest in those people. It is tempting to consider Fabian here: his assertion that anthropology often involves the assumption that the other exists in a different time has a particular resonance when the foreign other, in this case nineteenth-century Neopolitans, does actually live amidst a site so visibly associated with the past. Fabian considers the object-other, but the nineteenth-century Pompeians do not seem to have been any one's object, so the demand for a coeval view of another culture is not terribly apropos. In fact, Victorian visitors were not interested in asserting temporal and cultural distance from Pompeians past and present; they seemed rather to want to join those people in the past. Capturing this peculiar sense that past and present Pompeians live together at a site simultaneously ruined and inhabited, William Gell (1777–1836), famous for *Pompeiana* (1832), depicts original Pompeians alongside nineteenth-century visitors. In his "View of the Court of the Piscina," he places contemporary visitors at the center of the sketch and flanks them with detailed and vibrant portraits of Roman inhabitants of the city (see figure 4.2), bringing two distinct times together at this site and in this drawing.

Some visitors made a sport of inhabiting Pompeii and bringing the past to life. The author of "Pompeii by Torchlight" describes his experience dining with friends at Pompeii. He remarks on the absurdity of "respectable gentlemen in long-tailed coats, boots, and beaver hats" (66) supping in the same hall where long-dead Pompeian nobles would have eaten. The Romantic excursion, dining by candlelight amidst the picturesque ruins, worked on the imaginations of the participants to bring back to life those who dined there last. As they took their seats in the hall, the contemporary diners assumed the roles of Pompeians, reanimating the long-silenced dining room. The noise of their conversation and the clink of cutlery brought echoes of the past into the present. Two years later, another group played at reanimating Pompeii. In 1837, a person identified by the *Times* only as "an English gentleman" lodged for two weeks at Pompeii with his family and servants (20 October). The *Times* reports they dressed in Roman apparel and spent their time reading the classics. These reanimations differ markedly from the more accessible reenactments available at museums, where visitors walked through historical displays ranging from series of objects to full-scale re-creations of temporally and geographically distant places: at Pompeii, the re-creations were more interactive and involved reanimating the actual place by repeopling it. Museums, however impressive, were artificial in a way Pompeii seemed not to be. Furthermore, museums are public: in contrast, re-creations at Pompeii summoned domestic noises and sensations. No one reenacted a play or a gladiatorial contest in the arena; rather, the Victorians attempted to breathe life into Pompeian homes. Whether through an evocative piece of historical fiction or through an actual evening around a Pompeian hearth, nineteenth-century visitors and writers crafted a private sphere at Pompeii. Returning again to Fabian, it should be clear that what we see in the Pompeii I describe is an almost gleeful assumption of coevalness. The Victorians did not assume they were the same as the Pompeians; rather, they assumed the Pompeians were the same as them.[24]

Fig. 4.2. "View of the Court of the Piscina," William Gell's *Pompeii: Its Destruction and Re-Discovery* (1880).

Even Bulwer-Lytton repeatedly interrupts his historical narrative to describe nineteenth-century Pompeii. He guides the reader through broken houses, boasting about his reconstruction of the petrified city. In his introduction to *The Last Days of Pompeii*, Charles Dwoskin notes that even as late as the 1930s attendants at Pompeii recommended the novel as the best available guidebook to the ruins (ii). Particularly in his description of Glaucus's house, which he modeled on the so-called House of the Tragic Poet, Bulwer-Lytton indulges in tangents about the excavations and how the reader might find the ruins today if he or she should visit the site. He writes, "You enter by a long and narrow vestibule, on the floor of which is the image of a dog in Mosaic, with the well-known 'Cave Canem,' or 'Beware the dog'" (15). The tour continues: "Advancing up the vestibule you enter an atrium, that when first discovered was rich in paintings. . . . You may see them now transplanted to the Neapolitan museum" (16). Bulwer-Lytton continues on through the villa, even directing the reader to consult Gell's famous engravings of Pompeii, before returning to the story. By continually reminding his reader of the fragmented remains, Bulwer-Lytton shows off his ability to reconstruct those fragments into a vibrant city and narrative.

Bulwer-Lytton and many other writers celebrated the paradoxical conjunction of preservation and destruction. Pompeii was destroyed, but its ruins remain, and from the ruins, archaeologists reanimate the city of the dead. The point can be taken further: some admirers of Pompeii observed that it was only because of its destruction that Pompeii was preserved. In *The Buried Cities of Campania* (1868), W. H. Davenport Adams (1828–1891) claims that its tragic end secured Pompeii a far more notable place in history than it rightfully deserved. He paraphrases *Macbeth* in his preface: "Of Pompeii it may be said, that nothing in its history is equal in interest to its last scene. . . . this third-rate provincial town—the 'Brighton' or 'Scarborough' of the Roman patricians . . .—owes its celebrity to its very destruction" (v). Thomas Macaulay makes a similar gesture in his poem on Pompeii (1819). He describes the piecemeal demise Pompeii would probably have suffered if Vesuvius had not at once destroyed and preserved the city:

Had fate repriev'd thee till the frozen North
Pour'd in wild swarms its hoarded millions forth,
Till blazing cities mark'd where Albion trod,
Or Europe quak'd beneath the scourge of God,
No lasting wreath had grac'd thy fun'ral pall,
No fame redeem'd the horrors of the fall.
(119)

Macaulay contrasts Pompeii's perfect enshrinement with the gradual decay that befell most of the Roman Empire, a decay emphatically underscored by the erosion of words in the name of meter. Had fate spared Pompeii from sudden destruction, the city would have crumbled into oblivion as vandals carried away and scattered all its traces. Instead, Vesuvius froze Pompeii in its entirety, lifted it out of linear time, with each domestic detail in place for future generations to observe.

For popular writers and archaeologists alike, the tragic loss of life at Pompeii pales in comparison to the wonderful treasure trove of ancient relics excavated there. In "Preservation in Destruction," an 1852 article in *Household Words*, John Delaware Lewis describes the extraordinary value of the objects preserved by Pompeii's destruction and displayed at the Museo Burbonico. His hyperbole is typical:

> In the next cupboard are various articles of a domestic nature: soap, cotton, sponges, wax . . . colours laid out in readiness for the picture that never was to be painted. The vanity of human toil sowing where it is never to reap; the cunning of mankind intent upon a morrow which will never arrive; the value of small things; the worthlessness of great ones; —how many lessons are taught by these relics, the whole of which would not probably have purchased for their possessor a night's rest, or a meal, but for the possession of which the connoisseur would now-a-days mortgage his broad lands and entail poverty upon his unborn descendants? (283)

The original owners would never have saved these objects, but because they were incidentally preserved by the nature of Pompeii's destruction, they acquire a disproportionate value. Layard concludes his history of the excavations with the assertion, "We are thus indebted to Vesuvius for the preservation of the most perfect monument of the ancient world. The terrible mountain whilst it destroyed has also saved Pompeii" (1864, 348). Taking this point even further, the president of the Herculaneum Academy, quoted in "The Graffiti of Pompeii" published in the *Edinburgh Review*, "declared that, fearful as has been the ruin wrought by the successive eruptions of Mount Vesuvius, nevertheless the treasures of literature, science, and art which it has been the means of preserving to the modern world, more than compensate for the destruction brought on the victims of its fury" (411). While those victims might disagree, many writers suggest otherwise with their attraction to the rare memorialization of individual personalities and affections at Pompeii.[25]

In her short poem "The Image in Lava" (1828), Felicia Hemans (1793–1835) explores the complex relationship between preservation and destruction. Probably the most striking relics archaeologists uncovered at the site are the impressions left in the rock by fallen bodies. Vesuvius's ash acted as an extraordinary preservative, petrifying for posterity hundreds of Pompeii's citizens. The city offers no spectacular tomb commemorating its military or civil leaders; rather it displays the average person caught in his or her terribly private final moments. Layard describes the method of re-creating these bodies by pouring liquid plaster into the impressions (see figure 4.3):

> Amongst the first casts that [were] thus obtained were those of four human beings. They are now preserved in a room at Pompeii, and more ghastly and painful, yet deeply-interesting and touching objects, it is difficult to conceive. We have death itself moulded and cast—the very last struggle, the final agony brought before us. They tell their story with a horrible dramatic truth that no sculptor could ever reach. (332)

These casts epitomize Ricoeur's trace. Still on display at Pompeii today, the plaster casts capture the detail of sandal straps hurriedly tied, fingers clenched around prized possessions, and, most terrible of all, facial expressions. Hemans may never have seen these casts: her poem reflects on the fossil impressions of a mother and child who left a void in the rock at Herculaneum. She offers a sort of reverse ekphrasis, describing what is preserved only in its absence. She captures the vexed tone that underscores much of the literature on Pompeii as she paints a sentimental picture of the final moments of mother and child, yet she simultaneously grounds her portrayal in the brutal reality of a death so sudden and violent that the pair is forever petrified in stone. She reveals her preoccupation with the manner of preservation in phrases like "on ashes here impress'd" (22) and "cast in affection's mold" (36). Yet, Hemans's interpretation of these human fossils is laden with paradox. She writes:

> Temple and tower have moulder'd,
>   Empires from earth have pass'd,
> And woman's heart hath left a trace
>   Those glories to outlast!
>
> And childhood's fragile image,
>   Thus fearfully enshrined,
> Survives the proud memorials rear'd
>   By conquerors of mankind.
>   (5–12)

The "woman's heart" of line seven is an abstraction, a signifier of an intimate love, but at Pompeii, the phrase also suggests the rock cavity whose emptiness reminds the observer that a heart once stopped beating there. The heart leaves a physical "trace" that inspires sentimental imaginings. Hemans presents a similar paradox in lines nine and ten: "And childhood's fragile image, / Thus fearfully enshrined." Even with the awesome protection of a "woman's heart," this infant-child could not survive. Her death asserts her fragility, yet the nature of her shrine secures her lasting preservation. In death, the child's image "outlives" (11) whole civilizations.

Fig. 4.3. Photograph of plaster casts made from the impressions left by corpses at Pompeii. Photo by author, courtesy del Ministero per i Beni e le Attività Culturali-Soprintendenza Archeologica di Pompei. Unauthorized reproduction or duplication is prohibited.

Hemans not only emphasizes the preservation of the love and embrace between mother and child, but she also values the accidental monument to maternal love far above public memorials more easily erected and more often preserved. She contrasts "Woman's heart" and "childhood's fragile image" with "conquerors" and their "proud memorials." She reveres the private emotions of the home over the public triumphs of war and conquest. Moved by the love between mother and child, Hemans asserts the extraordinary value of these rare human relics:

> Oh! I could pass all relics
>   Left by the pomps of old,
> To gaze on this rude monument,
>   Cast in affection's mold.
>
> Love, human love! what art thou?
>   Thy print upon the dust
> Outlives the cities of renown
>   Wherein the mighty trust!
>   (33–40)

Again, Hemans sets "pomp" against "affection" and "human love" against mighty "cities." For her, Pompeii preserves something deeply human and far more precious than relics considered more grand.

If the authorities at Pompeii had prepared for their city's petrifaction, they would surely have brought public monuments, impressive artworks, and innovative architecture to the fore; they would probably not have bothered to display common crockery and children's toys. But the petrifaction was unplanned, and the relics left on display are simply the objects that were left behind. The accidental preservation of Pompeii resulted in the commemoration of the quotidian, and this glimpse into the private lives of the Pompeians captivated nineteenth-century visitors and readers. In the domestic relics at Pompeii, they discovered striking similarities to their own items of everyday use. Writing in *All the Year Round* (1884), the anonymous author of "A Last Day at Pompeii" explicitly likens life at Pompeii to nineteenth-century life:

> The coinage of that period differs little from our own . . . A gaming-table then was furnished with a pair of dice, and a lady's toilette-table with a mirror and a rouge-pot. Small boys scribbled on the walls, and played with balls . . . and marbles in the streets of Pompeii, as they do in modern Paris, Naples, London, or Berlin. . . . [T]here are traces at Pompeii of all sorts of London-shop things, and ways and means of living. (44)

Visitors were surprised, and surely pleased, to discover objects of daily life identical to those used in the present. This writer's diction even suggests that ancient Pompeii inherited its objects from Victorian London. A trace is a mark left in the present by something that happened in the past, but in this description, the mark of the present is evident in the past. This temporal conflation centers on the item of everyday use. Dice and rouge-pots become the fulcrum at which past and present swing together.

As they blend temporalities, Pompeii's broken walls and roofless houses also erode the sacred boundary between public and private. No one understood better than the Victorians the sanctity of the home, a refuge from urban miasma and the corruptions of public space. Indeed, protecting domestic space and ideals was of paramount importance in the nineteenth century. In their article "On the Parapets of Privacy," Karen Chase and Michael Levenson describe "the invention of a household universe" (426) and discuss the construction of external and internal walls to separate home and family from the threats of the street and even from the servants within the house. While Pompeian homes are more open, literally and figuratively, to the outside than Victorian homes, they nonetheless display efforts to protect the private space within. The mosaic floor in the vestibule of the House of the Tragic Poet depicts a fierce dog and warns, "*Cave Canem*," "Beware of the Dog." Clearly, the inhabitants hoped to keep out undesirable people and used the warning to secure some measure of privacy. As scores of visitors blithely walk across this once-effective barrier, they demonstrate not only how Vesuvius destroyed the life of Pompeii but how the eruption and the intervening centuries also wore down the parapets of privacy.[26] The private-made-public attracted nineteenth-century visitors who so valued their own domestic spaces and took a morbid delight in stepping through the ruined walls of Pompeii's private sphere.[27]

"A Day at Pompeii" revels in the opportunity to peek into the private spheres of the long-dead victims of Vesuvius. The author describes his visit to the city: "I dived into cellars, I ascended dilapidated staircases, I pried into ladies' boudoirs, nay, even their bed-chambers, stood before family altars, criticised the cook's department—in fine, explored with unblushing effrontery the domestic secrets of every household" (726). When tourists tramp through boudoirs, the private is made grotesquely public. This writer celebrates his effrontery and invites his readers and fellow visitors to join him in his invasion of Pompeii's privacy. Paradoxically, he appreciates the private because it is such a valued commodity in his own culture, but he fails to offer Pompeii's householders the same protection from intrusion that he would surely seek for himself and his neighbors. This author relishes his exploration of private spaces rarely preserved and almost never made available to visitors. The titillation of touring a lady's bedchamber is generally confined to the imaginative space of a novel, and the experience of a visitor to Pompeii, in this respect, parallels that of a reader of fiction. Both tourist and reader satisfy a voyeuristic desire to scrutinize the private spaces, thoughts, and actions of a stranger, and we might connect the attraction of Pompeii with the attraction of the novel.[28]

Perhaps the greatest writer of domestic fiction in the nineteenth century, Charles Dickens, not surprisingly, took an interest in the private remains at Pompeii. In *Pictures from Italy*, he describes the ruins of Pompeii as he saw them during his 1845 visit. He writes:

> Furniture, too, you see, of every kind—lamps, tables, couches; vessels for eating, drinking, and cooking; workmen's tools, surgical instruments, tickets for the theatre, pieces of money, personal ornaments, bunches of keys found clenched in the grasp of skeletons, helmets of guards and warriors; little household bells, yet musical with their old domestic tones. (1903, 171)

Like many of his contemporaries, Dickens notes the material goods that populated the everyday lives of everyday people in Pompeii, yet Dickens paints a more vivid picture than most. Note how alive the domestic remains are in contrast to the dead victims. The bunches of keys, reminiscent of Esther Summerson or other vibrant Victorian housekeepers, stand in stark contrast to the cold, impersonal grasp of the skeleton; the helmets have long outlived the guards and warriors they were built to shield; and the household bells still function, indicating a domesticity that survives even the most destructive catastrophe. Even when Dickens alludes to the spaces and institutions of public life, he trains his focus on the individual, private experience: he has nothing to say about the theater but notes the ticket stubs left behind by those who attended performances there, and he offers no general observation on the military but reflects on the helmets worn by particular soldiers. Dickens articulates what drew many to Pompeii's spectacular ruins: despite the havoc wrought by a volcano and the passage of nearly two millennia, the domestic perseveres. Of course, Pompeii has a forum and two theaters, several temples, and other public spaces. Archaeologists and writers of the Victorian period could as easily have focused on the public, but their own interest in the private directed their attention toward the relics of domestic, quotidian life. This emphasis reveals much about Victorian values, but it also reinforces the authority of the archaeologist to interpret his finds, of the historian to find the present in the past. The traces of life at Pompeii offer no clear narrative; those who examine those traces determine their meaning and are thus empowered to craft history. When they make that history familiar, they work to close the gap between past and present and thus lessen the impact of time's passage on individuals.

The unusual nature of its relics made Pompeii a fitting subject for a developing form of historiography. Historians' attention to material remains, instead of written accounts, became more frequent as archaeology and antiquities societies made available such elements of the historical record. In addition to altering the practice of history, these objects invited people of traditionally marginalized classes and professions into the historical narrative. While we consider recognition of this population a twentieth-century, postmodern development, the Victorians traced this brand of historiography to Niebuhr. Enacting his theory that the best historical studies are built from the detritus of popular culture, Niebuhr worked systematically with nontraditional, mainly literary, sources to develop a cogent narrative of Roman history. His approach seemed especially suited to Pompeii, with its remarkable wealth of cultural debris. Performing a sort of proto-cultural studies, many popular writers modeled their accounts of Pompeii on Niebuhr's history. Unlike Raymond Williams, who charts the development in this period of a concept of culture as an "abstraction and an absolute" (xviii), a collection of "certain moral and intellectual activities" separated from a new, industrial society (xviii), I argue that Pompeii reveals the Victorian interest in a form of culture that interests the postmodern reader. Pompeii offered a culture of lower- and middle-class men and women leading regular lives. For a historian interested in the relics of daily life, Pompeii is the sine qua non of archaeological finds. It presents the nonliterary but vividly telling texts of graffiti, ticket stubs, and an endless array of personal and domestic effects. From these objects, writers constructed compelling narratives describing life in Pompeii.

Among the most striking popular texts at Pompeii are the graffiti. "The Graffiti of Pompeii," published in the *Edinburgh Review* in 1859, describes the varied "wall-scribblings" found at Pompeii: school-boy slanders, lovers' memorials, grocery lists, and political slogans immortalized the simple, daily feelings of the real men and women of Pompeii (see figure 4.4). An 1881 article in *Chambers's Journal*, "Graffiti, or Wall-Scribblings," reprints a number of the more compelling examples: from domestic transactions—numbers of tunics sent to the wash or bunches of garlic sold—to derisive personal statements—"Oppius, ballet-dancer, thief and pilferer!" and "O Epaphras, thou art no tennis-player" (97). Like jewelry or keys, these relics preserve individual sentiments.[29] Echoing Niebuhr, the *Edinburgh Review* concludes, these "trifling things . . . serve better than many a more solemn, and seemingly more important, document, to realise to the imagination the men and things of the period to which they belong" (425). Anticipating contemporary Victorian cultural studies, this author asserts that graffiti offers insight into a portion of the population likely to be overlooked. He makes a surprising distinction between high and low art, when he likens the history reported by the graffiti to Henry Mayhew's *London Labour and the London Poor* and affirms the value of the low art represented by the graffiti. He participates in the mid-Victorian movement, famously championed by Mayhew himself but also by Dickens and, of course, by Engels and Marx, to extend attention and sympathy to the low and middle classes.

Fig. 4.4. Photograph of graffiti at Pompeii. Photo by author, courtesy del Ministero per i Beni e le Attività Culturali-Soprintendenza Archeologica di Pompei. Unauthorized reproduction or duplication is prohibited.

Understanding society means understanding all who comprise that society. Archaeologists at Pompeii share this interest. We might even hypothesize that Pompeii was interesting in part because it displayed the private lives of such a broad spectrum of society. The graffiti reveal Pompeii to the Victorian viewer with a realism impossible to glean through pristine statuary or public monuments. The *Edinburgh Review* reports:

> In the ardour of a first exploration, fragments like these were naturally neglected for what seemed to be of higher and more permanent importance; but it cannot be doubted that, rightly considered, they are not only extremely curious in themselves, but also calculated to throw light on the every-day life and manners of the ancient world, or at least to exhibit some of the lighter traits of popular character and the tone of mind which prevailed among a class to whose feelings and habits, as being unrepresented in the higher literature, hardly any other clue is now obtainable. (416)

While Pompeii was not the only ancient city to offer the careful excavator a glimpse at graffiti, it was the only site at which those private words complemented the other finds. Graffiti and personal ornaments made accessible the life of the average citizen at Pompeii, and such attention to the quotidian drew the Victorians to the city.

In *The Last Days of Pompeii*, Bulwer-Lytton offers the best-known and most extensive treatment of domestic life and personal relationships in Pompeii. He describes houses, personal effects, and dress, basing his fictional elaborations on the real relics uncovered at the site. He develops the character of Julia from his speculation about the body of an upper-class woman uncovered at the Villa of Diomedes; this is the same body described by Mariana Starke in Murray's 1832 *Travels in Europe*: "near the skeleton of this young woman were found several Necklaces, with other gold ornaments, silver and bronze rings, a piece of coral, a comb, &c, and in her hand, according to report, was a purse full of copper coins—perhaps, owing to the terror of the moment, mistaken for gold" (352). Bulwer-Lytton sets an emotional conversation between Julia and Nydia against the richly detailed backdrop of the lady's bedchamber:

> On the table before which [Julia] sat, was a small and circular mirror of the most polished steel; round which in precise order, were ranged the cosmetics and unguents, the perfumes and paints, the jewels and the combs, the ribbons and the gold pins, which were destined to add to the natural attractions of beauty the assistance of art and the capricious allurements of fashion. Through the dimness of the room glowed brightly the vivid and various colourings of the wall, in all the dazzling frescoes of Pompeian taste . . . Such was the dressing-room of a beauty eighteen centuries ago. (167–68)

He goes on to elaborate on the beauty's hairstyle, dress, make-up, and jewelry. Doing far more to fill in gaps than could any plaster cast, Bulwer-Lytton converts a human impression in rock into a rich ekphrastic description. He employs the same level of detail throughout the novel, and he almost always grounds his descriptions in actual relics. His enumeration of the delicacies served at Glaucus's bachelor dinner party includes many items excavated from the ruins, among them a Lesbian wine and pistachio nuts (22), and he lingers in his description of a particular goblet—"the handles of which were wrought with gems and twisted in the shape of serpents, the favorite fashion at Pompeii" (22). In this episode, as in many others, Bulwer-Lytton constructs from domestic relics a vivid illustration of daily life.

Though Julia dies, becoming the corpse Starke describes and perhaps also the owner of the boudoir so gleefully visited in "Pompeii by Torchlight," the central lovers of *The Last Days of Pompeii* do not die. With the help of the ever-faithful Nydia, Glaucus and Ione surmount the obstacles thrown in their path by jealousy and survive the volcanic eruption. In the ultimate triumph of the domestic, the novel's final chapter takes place ten years after the disaster in the almost Victorian home of the married couple. This last chapter is almost entirely a letter from Glaucus to his friend Sallust, datelined Athens. The epistolary form breaks down the barrier between past and present that the third-person narration maintained throughout the novel. The letter describes the happy life Glaucus shares with Ione: "Ione is by my side as I write: I lift my eyes, and meet her smile" (351). Finally, the two enjoy the Victorian ideal of a life lived side by side within the protective confines of a carefully constructed home. Glaucus describes his garden and his hall, spaces at the heart of the Roman home, which offer the inhabitants sanctuary from the crowded streets. In the end, the domestic ideal Glaucus and Ione embody survives, preserved by the volcanic debris of Pompeii and by the narrative authority of Bulwer-Lytton's historical fiction.

I will move now toward connecting London and Pompeii as archaeological sites, and I will begin with the export of Pompeii to London. Many designers incorporated elements or even exact duplicates of Pompeian homes in their London projects. Published in 1825, John Goldicutt's *Specimens of Ancient Decorations from Pompeii* was intended to provide suitable models for interior decoration. In 1844, Prince Albert commissioned a Pompeian room for the Garden Pavilion at Buckingham Palace; the room, destroyed in 1928, was an accurate reproduction of a room in a fine villa at Pompeii. Indeed, almost all reproductions of life and death at Pompeii emphasize the domestic objects and private spaces preserved at the site. The most impressive example of Pompeii re-created in Victorian London was at the Crystal Palace in Sydenham (see figure 4.5). The Pompeian Court was one of several courts, among them the Roman Court and the Nineveh Court, designed to give visitors to the Crystal Palace Park an opportunity to stroll through an impressive classical environment.[30] Interestingly, in his booklet on the Pompeian Court, George Scharf explains, "The original intention in constructing the Pompeian Court

Fig. 4.5. "Pompeian Court" at the Crystal Palace, Sydenham, courtesy Guildhall Library, City of London.

in the Crystal Palace was to appropriate it for purposes of refreshment.[31] In furtherance of this plan, more especial attention would have been devoted to the mural decorations and the arrangements for public accommodation and convenience" (43). Scharf thus reveals that while "each part [of the house] has been copied from some existing authority" (44), there was an object driving its design that was very modern. Not only did the Victorians interpret Pompeii on their own terms, but even when they reconstructed a bit of Pompeii in their own capital, their contemporary desires took precedence over historical authenticity. Scharf imagines that even the original Pompeians would have noted the difference; he speculates that they "would, ages ago, have found nothing strange and nothing amiss here, excepting the appearance of the thronging visitors . . ." (65). The very presence of visitors makes the site Victorian as opposed to Pompeian. It becomes a home on display, its private spaces thrown open to the public, and it thus becomes a museum piece rather than a true re-creation. However, Scharf defends the Pompeian Court as actually more realistic than the ruins at Pompeii. He observes that "visitors to Pompeii look for excellencies that do not exist, and a harmony incompatible with the actual condition of the remains; and, discontented at finding things in opposition to their own conceptions, they depart with imperfect and even prejudiced ideas of what they really have beheld" (72–73). In contrast, the visitor to the Pompeian Court views a perfect Roman villa, its walls and ceiling complete, its artwork unblemished. Thus, one visits a site in London that is closer to the reality of Pompeii than those sites that one may visit in Pompeii itself. Reconstruction triumphs over ruin, and the presence of "throngs" does not compromise the integrity of the visit. At the Pompeian Court, Victorian archaeologists crafted the perfect narrative of Pompeii: they reanimated its splendor, repeopled its homes, and made complete the transfer of Pompeii to London.

Digging into Londinium

Pompeii and London differ fundamentally as archaeological sites because Pompeii is, despite its few inhabitants and eager reenactors, a city of the dead, whereas nineteenth-century London was very much alive. Pompeii appeared whole, while Londinium was fragmented, offering only isolated artifacts or discrete ruins disconnected from other archaeological material. In terms of the archaeological landscape and the intrusion of the present day, the two cities are quite dissimilar; however, striking parallels were observed, even insisted upon, between the present and future of London and the ruin of Pompeii. In other words, the relics of Londinium signified little about the Roman Empire and much about nineteenth-century London.

The urban improvements that so characterized the landscape of Victorian London achieved not only progress toward a new and improved London but also an intriguing regress to London's imperial past. That is, the excavation of London in the name of progress unexpectedly uncovered the city's Roman origins. In this section, I will discuss urban archaeology, also known as rescue archaeology. This archaeology was almost always the by-product of works to improve the city. The extensive works to construct the Thames embankment, for instance, led to the discovery of many items from Roman London. Although these improvements are not themselves my subject,[32] a brief survey of the range of works in progress shows the wealth of excavatory sites incidentally available to archaeologists. Urban improvements in nineteenth-century London included the construction of sewers, the Thames embankment, and new roads. Building infrastructure required eradicating the old, and clearances of slums, old buildings, and roads paved the way for improved structures. The construction of London's underground is a good case in point.

Lynda Nead writes boldly, "the building of the Metropolitan Underground Railway wrecked London" (37). She goes on to describe a new urban aesthetic introduced by the excavation for the railway: she writes, "the tunneling itself summoned images of the sublime, with excavations on the apparently limitless scale and tiny figures dwarfed by massive building works. This was a new urban aesthetic built around the forms of the tunnel, the trench, the vault and scaffolding" (39). Strikingly, this is also the aesthetic of the archaeological excavation. Figure 4.6, an image from the *Illustrated London News* of 1862, shows the construction of the railway. The area of the works is clearly extensive, and the contrast between the scaffolded ditches in the foreground and the buildings in the background suggests destruction—similar buildings must have stood where the construction now consumes the foreground. Workers are pictured scattered throughout the scene, laboring to complete the work and bring progress to London, although, as Nead observes, the workers are dwarfed by the works. The area is a morass of ditch and scaffold, the underside of London revealed but then harnessed by the apparatus of improvements. An image like this one shows the progress toward improvement, but it cannot conceal the destruction inevitable en route. To build a sewer, a road, or a railway, buildings and streets were demolished. These works in progress gave mid-century London the appearance of a city in ruin.

Nancy Metz suggests that the improvements offered Londoners a vision of their city as it would be in the future, a ruin no different from Pompeii. She writes,

Though these excavations were carried out by engineers and city planners rather than "adventurous Belzonis,"[33] even such modern improvements as the building of railroads, the laying of sewers, and the demolition of slums brought before residents of the capital an increased awareness of London as an archaeological artifact. . . . metropolitan improvements wrought, visibly, instant "ruin" on whole neighborhoods. (1990, 476)

Fig. 4.6. "The Metropolitan Railway and the Fleet Ditch," *Illustrated London News* (1862).

Archaeology resonated with Victorian Londoners who knew all too well demolished streets, scaffolded monuments, and ruined houses. They looked at their capital and could not readily discern whether they saw a great city under construction or whether they were witness to its destruction. This simultaneous progress and decay resonated also with vexed responses to London and the very project of urban reform. Many wondered if improvements would truly improve city life, and several writers expressed nostalgia for the picturesque, unreformed city of old. Nead asserts that the *Illustrated London News* took up the banner of the picturesque, reporting on the demolition of old buildings: "In regular columns with titles such as 'Nooks and Corners of Old England' and 'Archaeology of the Month,' it illustrated the disappearing inns and houses of Elizabethan London, creating in its pages a lexicon of the metropolitan picturesque" (31). Perhaps so-called progress was, in fact, destructive.

The view that progress might be regress was confirmed by the unplanned partnership between metropolitan improvements and urban archaeology. Excavation in the name of the future led quite literally to London's past, and, in this sense, progress and ruin become intriguingly synonymous. With its trenches and scaffolds, London illustrated the temporal coevalness I have discussed. Past, present, and future seemed to coexist in the singular moment of the nineteenth-century city. In the quagmire of urban improvement, Londinium asserted itself. Discoveries of Roman remains were everywhere, and the reports of these discoveries became commonplace in popular publications like the *Times*, the *Illustrated London News*, and *The Gentleman's Magazine*, as well as more specialized journals such as *Archaeologia*. Digging the foundation of the Coal Exchange, workers discovered a Roman villa in Lower Thames Street, a Roman gravel pit was found beneath the Royal Exchange, workmen uncovered mosaic pavements in Threadneedle and Bishopsgate Streets (among many others), and assorted relics of Londinium—coins, sandals, and even bronze sculpture—were found in the Thames. In the first volume of *The Archaeological Journal* (1888), John Edward Price (d. 1892) compiles an index of all the Roman artifacts discovered in London. Ranging from spoons, bottles, and hairpins to mosaic pavements and partial dwellings, these artifacts were discovered at 208 distinct locations throughout London. In an 1866 review of London excavations published in *Archaeologia*, William Henry Black (1808–1872) recounts his accidental discovery of a Roman foundation on St. Peter's Hill: ". . . I observed the workmen belonging to the City Sewers department excavating the ground for drainage, and casting up portions of Roman brick and concrete. I immediately caught up a portion of that brick . . ." (48). Black's diction suggests Roman relics fly haphazardly about the city, the past hurtling into the present. He further suggests his own importance as an archaeologist, supervising the workmen and deftly catching an artifact as it careens carelessly into the modern city.

Rescue archaeology describes the primary sort of recovery work undertaken in London in the nineteenth century. Charles Roach Smith (1807–1890) was the most renowned figure in London archaeology; he collected relics throughout the city and wrote extensively on the nature of his finds. Though few now know Roach Smith by name, many have seen some of the artifacts he recovered from London's depths and which formed the core of the British Museum's Anglo-Roman collection.[34] Roach Smith was trained as a chemist and developed an interest in archaeology during his residence in the City of London. He observed that in the excavation accompanying so many of the city improvements, workers routinely uncovered a variety of Roman artifacts and even substantial structures. He intervened in these city projects and did his best to salvage these accidental archaeological finds.[35] By the 1840s, Roach Smith was recognized as the leading expert on Londinium, and although he was repeatedly frustrated in his attempts to ensure the preservation of ruins in their original location, he did a great deal to make the remains accessible to an increasingly interested public. When a private collector offered him 3000 pounds for his Roman artifacts, he declined, preferring to keep the collection whole and available for display in a public space. In 1856, the British Museum gave him 2000 pounds for his collection, and, in the bargain, he also acquired enough recognition to publish by subscription his *Illustrations of Roman London* (1859).

A review of Roach Smith's book stresses the importance of the finds and even makes explicit the connection between London and Pompeii:

> We are now reaping the fruits of that philosophy of history which was inaugurated in the last century, and which has ever since been dissipating or restoring our traditions. We have just acquired the conviction that Roman London was once a fact, and, as far as we can now judge, a great fact. Every schoolboy knew that it was to be found in the ancient atlas, but even the most studious persons were obliged to take its importance upon trust, as a place of which a very few monuments survived to tell of its site, extent, or pretensions. How much there was of this Roman city to be still detected and disinterred from beneath our cellars and streets was never known adequately till this volume of Mr. Roach Smith set its principal relics in splendid profusion before the eyes of the elect. We discern in the *Illustrations of Roman London* by this able antiquary an epoch in the discovery of the city under our feet—a city which as certainly lies there in some sense as Pompeii under its ashes. (*Times*, October 19, 1859)

This reviewer credits Roach Smith with making real what had seemed fantastic. He suggests that without the archaeological evidence discovered and described by Roach Smith, Londinium would not exist.

Yet, Roach Smith acknowledges in *Illustrations of Roman London* the accidental nature of the work that led to his archaeological finds: "The excavations, which led to those researches, were made for sewerage, for what is commonly termed 'city improvements,' and for deepening the bed of the Thames to facilitate navigation" (i). A few years earlier, in the introduction to his *Catalogue of the Museum of London Antiquities* (1854), he was even more explicit: "The collection has been formed under circumstances entirely accidental" (iii). Roach Smith purchased some artifacts from workmen after hearing of their incidental discoveries; others he acquired through archaeological detective work. For instance, knowing that a Roman gravel pit had been discovered beneath the Royal Exchange, he carefully observed other works in the vicinity, and in this manner, he managed to poach the labor of the city workers, turning their digging for improvements into digging for archaeological discovery.

For Roach Smith, archaeological discoveries were inseparable from the metropolitan works that brought them to light. A piece titled "Roman London" in *Bentley's Miscellany* from 1866 describes the digging that was the source of Roach Smith's collection and the increasing wealth of antiquities held by the British Museum and others: "Mr. C Roach Smith's museum contained over five hundred relics of Roman London, collected in the metropolis during street improvements, sewerage, and the deepening of the bed of the Thames, and many additions have since been made from the same sources, to which are to be added the objects discovered during the extensive clearances effected for railways" (366). Writing in *Archaeolgia* in 1840, Roach Smith explains, "I am enabled to exhibit to the Society of Antiquaries some fine and interesting bronzes, which were found in the bed of the river Thames, near London Bridge, in January last, by men employed in ballast heaving" (28). Two years later, Roach Smith describes the "perfect tessellated flooring of a room" along with walls and passages "furnished during the progress of excavations for the south wing of St. Thomas's Hospital" (29). He discovered a tessellated pavement "in 1854 between Bishopgate Street and Broad Street when the Excise Office was pulled down" (1859, 54). The illustration shown in figure 4.7 is striking because it depicts the evidence of the works in progress, with the shovel and pickax in the lower left-hand corner, but it shows also a gentleman, perhaps Roach Smith himself, inspecting the find. Like Lyell's geologist sketching the Temple of Serapis, the archaeologist in the margins reminds the viewer that these remains only have meaning once they've been excavated and interpreted. Yet, the meaning uncovered through urban archaeology is in danger as the drive to improve the city threatens to obliterate traces of the past even as it thrusts them into the light of the present day.

As the force behind these improvements, the City Corporation is repeatedly represented as Roach Smith's nemesis, openly hostile to preservation of antiquities. Even stopping work long enough to allow someone to document a wall or a floor before digging through it meant halting progress, and the "corporation was not embarrassed by any such solicitude" (*Illustrations* ii). Writing in 1842, Roach Smith complained, "No provision is made by the body corporate of our venerable city to turn aside the axe and spade at the approach to a frescoed wall, or a rich mosaic pavement. The hand of unchecked ignorance

Within the image: Drawn & Engraved by J.W. Fairholt

March 1854

Fig. 4.7. "Tessellated Pavement," Charles Roach Smith's *Illustrations of Roman London* (1859).

in a few minutes destroys what time has spared, and often before it is possible for the antiquary to make even a memorandum of the fact."[36] He goes on to attack the improvements directly: "In a very few years, the very vestiges of [Londinium's] walls will be swept away by the ruthless hand of *soi disant* improvement" (1842, 150). The opposition from the City Corporation compounded the always-difficult work of urban archaeology and required heroism and sometimes stealth from the city's archaeologists. Writing in the *Archaeological Journal*, Roach Smith describes the many obstacles to excavation:

> It must be obvious to all who consider the present condition of the metropolis of England, that great difficulties would beset any attempt to carry on a systematic exploration of the wreck and ruins of the ancient town, buried beneath the accumulated soil of centuries and the crowded masses of modern buildings. Under the most favourable circumstances such a project would encounter objections almost insurmountable; but when undertaken by individual zeal on a partial and confined scale, at uncertain times and places, whenever the earth may be excavated for public works, without assistance or countenance from the directors, and usually in contention with obstructions and annoyances of all kinds, it is fortunate, in such a state of things, should any discoveries be rendered available to the topographer and antiquary. (108)

Buried in this diatribe is the assertion that the excavation of Londinium is only possible because of "individual zeal," the persistence of archaeologists committed to preserving the past.

However loathe he may have been to admit it, Roach Smith understood that the progress toward London's improved future led simultaneously, if incidentally, to Londinium's buried past.[37] Though he was highly critical of the City Corporation, he could not have recovered any remains of Roman London without the excavations sponsored by the City. And those writing on behalf of the City readily acknowledged the connection between improvements and archaeology. William Tite (1798–1873), the architect of the New Royal Exchange, responded directly to Smith's accusations of destructive indifference in 1848. He suggests the Roman remains are especially valuable *because* they are the by-products of city improvements:

> This increased intimacy with the nature and value of antiquities has led to their more careful preservation and better exhibition, as well in public local depositories as in private collections. One of the former is the Museum established in connection with the Corporation Library at Guildhall, for the reception of antiquities relating to London, especially such as may be discovered in execution of civic public improvements. (37)

Tite does not elaborate on why those remains discovered through metropolitan improvements should be especially noteworthy, but he may have perceived the value of anachronistic space thus revealed: London's imperial strength is implicitly connected to its geographic past—that is, Rome's bygone imperial strength and eventual decline. I should point out here a more complicated diachrony than I have yet allowed: "past" in London signifies not just a particular moment in Londinium (in contrast to Pompeii, when the past preserved is a specific day) but all moments—moments when the Roman Empire was at its height and later moments of decline. The observer chooses which past time to evoke, and it is the coevalness of all times sharing the same physical space that facilitates such selective diachrony

In his "Theses on the Philosophy of History," Walter Benjamin muses on the relationship between past and present in historical discourse. His reflections offer a philosophical explanation for the ease with which the Victorians saw the connection between Londinium and their own nineteenth-century London. In his fifth thesis, Benjamin writes, "for every image of the past that is not recognized by the present as one of its own concerns threatens to disappear irretrievably" (257).[38] This assertion that the past is only interesting insofar as it reflects the present builds on other claims Benjamin makes, such as his often-quoted statement from the seventh thesis: "Whoever has emerged victorious participates to this day in the triumphal precession in which the present rulers step over those who are lying prostrate" (258). The seventh thesis is of special interest to historians of archaeology because it addresses the function of artifacts in the construction of historical narrative and contemporary political power; Benjamin goes on to mention the "spoils" carried along by the victors, and thus evokes not only historical curiosities but also the imperial ideology that co-opts them. At a most basic level, Benjamin describes a sort of temporal relativism in which the past is unavoidably viewed through the lens of the present and is thus defined not simply by the concerns of the present but by the people in the present who wield cultural power. Central to the relativism Benjamin describes is the notion that the past is fragile. At a literal level, we might think here of the relics crushed by Belzoni's boots or turned to dust through Layard's careless excavation; in both instances the force of the nineteenth-century archaeologist crushes the remains of the past. At a metaphoric level, we should recall the description of the geologist from chapter 2: the past requires an observer, without whom artifacts are meaningless and slip away into oblivion. The trace—the entirety of the past—depends upon the present for its continued existence.

The accidental archaeology of nineteenth-century London provides a scenario that further complicates Benjamin's description of the relationship of past to present. He claims that the present must recognize the past as one of its concerns. He elaborates in the sixth thesis: "historical materialism wishes to retain that image of the past which unexpectedly appears to man singled out by history at a moment of danger" (257). Regarding their sense of their relationship to history, the Victorians were, it is fair to say, in a "moment of danger." Recall Tennyson's infant crying in the night, evocative of the desperation and isolation so many felt in the face of the geological time scale and all its implications. At this particular moment of danger, the Victorians questioned their identity not only as a culture and as a nation, but more fundamentally as a species. Artifactual evidence of dead cultures spoke to them with an urgency that archaeology rose to address. Yet, not all in the present recognized Roman remains as their concern. Many of the workers, and those who employed them, had an eye only on the future, the end of the progress in which they were direct participants. It took an excavator like Charles Roach Smith to see the profound temporal analogy: as Londinium is to London, so London is to some future state. The present is the moment of danger, and those who exist in the present must look at once to the past and to the future to rescue themselves.

The daily excavation of London for the purpose of urban improvements rendered the site an eerie ghost of its future self as, like Pompeii, an archaeological relic. Speculation about the future of London ranged from the apocalyptic to the humorous, from the delightful to the morbid. In one of the best-known descriptions of London in ruins, Macaulay imagines a future "when some traveler from New Zealand shall, in the midst of a vast solitude, take his stand on a broken arch of London Bridge to sketch the ruins of St. Paul's" (1840, 228). Many writers throughout the nineteenth century refer to Macaulay's New Zealander: notably, Blanchard Jerrold (1826–1884) and Gustave Doré (1832–1883) conclude *London: A Pilgrimage* (1872) with a striking illustration of London in ruins, the New Zealander sketching the dome of St. Paul's amidst the rubble (see figure 4.8). Six years after Macaulay's New Zealander penetrated the popular imagination, "London in A.D. 2346," a satiric piece in *Punch*, provocatively likened London's future ruin to Pompeii's, even attributing the destruction of London to a Mount Vesuvius. The article reports the proceedings of a meeting of the New London Archaeological Institute in April of the year 2346. Following the style of many articles on Pompeii, the author takes the reader on a tour through the ruined city, Trafalgar Square replacing the Forum. The author meanders through the National Gallery, noting the poor quality of the art and the "dank" (186) display areas. When he exits the gallery, he puzzles over the fountains in the square, theorizing that the basins proffered soup to the poor. He writes, "The utter want of proportion between the basins and the centerpieces, and the miserable dimensions of the latter, point rather to some purposes of a domestic or culinary character, in which ornament has been sacrificed to use" (186). The author makes the valuable point that surely some conclusions about Pompeii are false, based on contemporary tastes rather than an accurate understanding of the ancient culture. He even implies an imposition of a domestic ideology on what might actually have been an array of public monuments. Nonetheless, while it makes light of archaeological interpretations, this article seriously ponders London's legacy.

Fig. 4.8. "The New Zealander," Blanchard Jerrold and Gustav Doré's *London: A Pilgrimage* (1872).

Other writers imagine a terrifying, disinherited future in which no trace remains of the present life. Richard Jefferies (1848–1887) offers an extended speculation on the future of London in his dark and disturbing *After London* (1885). Jefferies describes a London reclaimed by nature, not unlike the unexcavated Pompeii, but Jefferies's eruption is gradual and verdant rather than sudden and fiery. Most disturbingly, his dead city leaves no trace of ever having lived. The greater power of nature is asserted and the city made to seem insignificant, even frail, in the face of time and undergrowth. Jefferies writes:

> For this marvelous city, of which such legends are related, was after all only of brick, and when the ivy grew over and trees and shrubs sprang up, and, lastly, the waters underneath burst in, the huge metropolis was soon overthrown. At this day all those parts which were built upon low ground are marshes and swamps. Those houses that were upon high ground were, of course, like the other towns, ransacked of all they contained by the remnant that was left; the iron, too, was extracted. Trees growing up by them in time cracked the walls, and they fell in. Trees and bushes covered them; ivy and nettles concealed the crumbling masses of brick. (48–49)

Brick is no match for nettles, and so London is consumed by the very ground it once dominated. In Jefferies's future London, the city diminished to the faintest of memories, the people have all become "bushmen[,] . . . gypsies . . . and half breeds" (26ff)—even the English race has eroded.

This London of disintegration and devolution nagged at a public preoccupied with the forces of destruction and preservation. Indeed, Jefferies's late-century book exemplifies post-Darwinian concerns about degeneration.[39] *After London* is subtitled *Wild England* and thus explicitly imagines a fall from civilization and return to wilderness. Fears that humans and their societies might degenerate rose directly out of Darwin's *The Origin of Species*: if humans were biologically linked to apes, and if they had evolved from that state, then couldn't they as easily devolve and return to their origins as wild primates? Colonial expansion and exposure to so-called primitives seemed to confirm the possibility that humans might exist in a lower state. This notion was applied to England, McClintock argues, where the working classes and women were viewed as atavistic—that is, primitive and degenerate. Tennyson addresses the subject of degeneration explicitly in "Locksley Hall Sixty Years After": "Evolution ever climbing after some ideal good, / And Reversion ever dragging Evolution in the mud" (197–98). He is more clear about the fate of humans earlier in the poem, when he asks, "Have we risen from out the beast, then back into the beast again?" (148). Troubled by the implications of evolution and its imagined partner degeneration, as well as by the evident origins of humans as prehistoric cohabitants with mammoths and the historical certainty that empires end, the Victorians sought the answer to the question posed boldly in the title of an 1852 *Household Words* article, "What Is to Become of Us?"

Pompeii's monumental petrifaction assuaged the anxieties implicit in this question. London need never be removed or consumed by hungry weeds; rather, it might become a Pompeii-like relic and thus through preservation evade degeneration. With the strategic application of the science of archaeology, the transformation into a relic city and the future interpretations of its traces seem manageable. In "A Last Day at Pompeii" (1884), an author asks: "If the Downs became volcanic and Brighton were entombed, what portion of its statuary could be deemed worth preservation for eighteen hundred years?" (45). Other writers and planners answered the question by proposing careful and selective preservation of the present. Proposing a plan for preservation in an 1890 article, "A Pompeii for the Twenty-Ninth Century," the positivist Frederic Harrison (1831–1923) takes Pompeii as his model as he presents an elaborate project intended to guarantee the immortality of the Victorian age.[40] While Harrison's meticulous plan does not perfectly reflect the historical outlook of his contemporaries, who might have chosen an evolutionary display showing the Victorian culture as the pinnacle of civilization, many did share his concern over what sort of trace the present can offer the future. When they asked, "what is to become of us?" the Victorians wondered not only what would remain of their culture, but also how their remains would be interpreted in the future. Noting "we live in an age of archaeological research," Harrison chastises his contemporaries for their absorption in the past at the expense of the future and writes: "Let us . . . take to *looking forwards*; and, with all our archaeological experience and all the resources of science, deliberately prepare a Pompeii . . . for the students of the twenty-ninth century" (Harrison's emphasis, 381).[41] Harrison offers one moment in time from one culture frozen forever. His time capsule might be akin to McClintock's anachronistic space, although it is certainly a space created by and largely celebrating the ruling classes. What is particularly striking about Harrison's argument in the context of urban archaeology is that he takes the focus on the future associated with public works and transforms it into an archaeological perspective.

Harrison bemoans the loss of invaluable objects through neglect and misinterpretation; he cites the "brutal ignorance of man, the blear-eyed stupidity of monks, the ambition of kings, the greed of traders, and the slow all-consuming dust of ages" (382). He suggests a plan that will ensure the preservation of the artifacts produced in his own age:

> Let the science and learning of the nineteenth century do for the twenty-ninth century what we would give millions sterling to buy, if the ninth century AD, or the ninth century BC, had been able and willing to do it for us. In other words, let us deliberately, with all the resources of modern science, and by utilising all its wonderful instruments, pre-

pare for future ages a sort of Pompeii . . . wherein the language, the literature, the science, the art, the life, the manners, the appearance of our own age and its best representatives may be treasured up as a sacred deposit for the instruction of our distant descendants. (383)

Note Harrison's emphasis on the deliberate quality of this monument, intended to immortalize the Victorian present. Harrison assumes the authority not only to narrate the histories suggested by ruins, but also to deliberately craft remains.

While Harrison claims to look only to the future, his emphasis is actually on the present. He insists, "Let us have a rest from Great Exhibitions for a year or two: and try to organize a posthumous Exhibition for the benefit of posterity" (389). But, of course, even as it benefits the archaeologists of the future who will open the vaults, Harrison's plan benefits the people of the present who will achieve immortality through the deliberate Pompeii. In fact, Harrison argues that the temptation of self-preservation will attract many notable persons to the project. He asserts that state and church would be unable to agree on the objects to include and so proposes that the project be funded by voluntary effort, and especially from the free donation of valuable works by the authors themselves. This self-preservation would surely be attractive to figures who desire immortality for their art: Harrison writes, "The difficulty of the committee of selection would be to refuse, to reject, to exclude. Artists, authors, inventors, and producers of all kinds would be only too eager to deposit works which would be destined to so distant and certain an immortality" (388–89). Clearly preservation of the present is as important as offering historical data to the future.

Harrison devotes the greater part of his article to enumerating the objects he feels would best represent the Victorian Age. Although he mentions generally "the life, the manners, [and] the appearance" (383) of the period, he specifies primarily great works and celebrated individuals. He proposes an impressive array of cultural objects, among them photographs of statesmen and a phonographic recording of Tennyson reciting *The Princess*, which, he writes, "it would be the business of Mr. Edison to protect for a thousand years" (386).[42] Also protected should be copies of the *Encyclopedia Britannica*, the *Times*, and the British Museum *Catalogue* in addition to "models of a locomotive . . . and of the House of Commons . . . along with a dressed model representing Mr. Irving in Hamlet, and a fine lady dressed for a drawing-room" (386). These models are a far less terrifying version of Pompeii's plaster casts: they show people living their lives, but they do not so adamantly signify the ends of those lives. Harrison also insists on including an "electro-photographic reprint of the Ordnance plan" (388) of London for the benefit of Macaulay's New Zealander and others who visit future-London.

Yet, alongside these acclaimed works and representative guides, Harrison champions the value of noncanonical texts and images. Much like Pompeii's graffiti, certain curiosities may paint a richer picture of the age. In defense of preserving transcripts of Cockney dialect and works of popular fiction, he asserts: "Our curious young New Zealander . . . would no doubt much prefer a Whitaker's *Almanack* or a Bradshaw's *Railway Guide* of 1890 to all the works of Mr. Froude or Robert Browning. Which would we rather have today—the epics of Lucius Varius, or a complete gazetteer, or post-office directory, of Rome under Augustus? These things should not be left to chance" (388). Harrison esteems the items that record quotidian life and plans for the inclusion of such objects in his vault. The scope of the potential relics Harrison enumerates reflects the historiography that took special interest in Pompeii: material goods tell stories and penetrate social boundaries. Interestingly, Harrison insists upon the value of the very texts that twenty-first–century scholars have begun to study and integrate into the general understanding of the Victorian age. Fortunately for Harrison, whose Pompeii for the twenty-ninth century never was constructed, chance, in the form of Victorian cultural studies, has after all preserved the texts he extols.

Like Pompeii, Harrison's extraordinary monument includes a range of representative objects from all levels of society, but it differs significantly from the original in being deliberately constructed.[43] A product of bureaucracy and technology, the English Pompeii will preserve only those items submitted by hopeful artists and then vetted and selected by a committee. Thus, the future Pompeii fails to capture the private life the real Pompeii so poignantly memorializes. The past Pompeii accidentally preserves the daily items in use when the tragedy struck. There was no manipulation of the culture, no leaving out the lesser painters or hiding away the cracked pots. Pompeii attracted the Victorians largely because of this spontaneity. Harrison's Pompeii carefully nested in Salisbury Plain would monumentalize a range of Victorian products, from poetry to political pomp, but it would not offer the future-archaeologist access to a middle-class pantry or shopping list, and it would certainly not preserve the fallen figures of a mother and child.

Nor would any deliberate Pompeii preserve the unfinished works that revealed how the real Pompeii was so tragically cut off in the midst of life. Several descriptions of the site call attention to the repairs underway in the city when it was petrified. In "The City of Sudden Death," Lewis writes, "As we return through the ruins of the Stately Forum, let me call your attention to these fragments of columns lying on the ground—or rather masses of stone, half-worked into the shape of columns—the final catastrophe having come at a time when the Forum was itself under repair. Do you see that last mark of the chisel? Do you notice where the fluting has been abruptly left off?" (176). An earthquake struck Pompeii in 62 CE, and all around the city archaeologists found evidence of the reconstruction in progress when the volcanic eruption stopped all work. In "Latest News from the Dead," published in *All the Year Round* in 1863, the author observes, "as we walk along its streets, we not only

see the theatre and many other edifices to have been in the process of reconstruction at the time of their burial, but, in the quarter once occupied by the
stone and marble masons, there lie portions of an old frieze . . . beside which
stand copies of the same decoration cut in white marble ready for erection in
a restored temple" (473). These images struck nineteenth-century viewers
who were moved by this evidence of common people blithely progressing
through their daily lives when destruction came suddenly upon them. The
signs of urban improvement also brought to mind another city: London was
similarly a site of decay and progress, of excavation that revealed the past as it
led to the future.

From the fragments discovered at Pompeii and London, nineteenth-
century writers reconstructed the past in the image of their present and with an
eye to their future. Archaeological remains offered the possibility of cultural
immortality, an inviting stasis: perhaps in two thousand years, an excavator might unearth a nineteenth-century map of London, and a poet might
repeople the fragmented city. Yet, despite her varied imaginings, no poet or
positivist could really prepare a Pompeii for the future. Pompeii preserves the
private as well as it does only because it was unprepared, and even the preservation of the city is, in the final analysis, inexact. The private is made perversely
public, and the whole is interpreted with the bias of observers who place a pre-
mium on the domestic. In "The Archaeology of Victorian Literature," Steve
Dillon writes, "Victorian writers followed the geological dismantling of Genesis and the apocalypse of the French revolution, and their thoughts dwelled on
the difficulty of discerning origins and ends and on the profound sensibility of
following" (Dillon's emphasis; 243). I would add that their thoughts dwelled as
much on the sensibility of remaining: nineteenth-century historians began to
realize that historical narratives were constructed from imperfect evidence, notably the material debris of popular culture, and their new historiography
begged the question, "what is to become of us?" Even "us" is telling diction:
the question asks not what will become of a nation or a species but what will
become of a collective of individuals. The archaeological sites of Pompeii and
London offered an answer: the individual need not decay and vanish over time;
rather all times might be coeval and the very forces that destroy might also preserve. Thus reconciling themselves to the immensity of time, early Victorian
archaeologists rescued the individual from ruin.

5

Dickens
among the Ruins

Charles Dickens's *Bleak House* opens with an image of geological time oddly yet familiarly imposed on nineteenth-century London:

> London. Michaelmas term lately over, and the Lord Chancellor sitting in Lincoln's Inn Hall. Implacable November weather. As much mud in the streets, as if the waters had but newly retired from the face of the earth, and it would not be wonderful to meet a Megalosaurus, forty feet long or so, waddling like an elephantine lizard up Holborn Hill. (49)

As a dinosaur waddles through Dickens's London, so do all sorts of artifacts, ruins, and traces of the past corrupt the experience of living in the present and render stagnant the hope of progress toward the future.

It is not at all surprising to find a prehistoric beast in Dickens's idiom for he, like his contemporaries, could not avoid geology's compelling evidence of different worlds that existed before our own. Nor is it at all surprising that he places that beast in mid-Victorian London: like Thomas Hardy's Henry Knight who meets a trilobite in *A Pair of Blue Eyes*, Dickens's characters must stare down geologic and cultural pasts, as well as so many personal ones. Like the city workers who accidentally became archaeologists, suddenly uncovering a Roman wall as they dug a railbed, Dickens's characters cannot escape the ruins that assert themselves into the present urban landscape. Like geologists or archaeologists, they reassemble ruined fragments into whole narratives—

Amy Dorrit and John Rokesmith both use ruin and excavation to restore their identities. Dickens offers excavation and complementary images from archaeology and geology as narrative tools: recovering fragments from the past and crafting from them a story in the present asserts the value of the individual in an indifferent world. This chapter examines Dickens's use of motifs taken from geology and archaeology in *Little Dorrit* (1855) and *Our Mutual Friend* (1864). I argue that Dickens uses literal and metaphoric excavation in both novels to chart his characters' relationships to the past and to equip them to salvage their own worth from the depths of the past. In *Little Dorrit*, he represents the oppression of time in a post-geology world but concludes with an emphasis on gradualism that describes the possibility of escaping time's indifferent expanse. In *Our Mutual Friend*, he is less interested in historic time, but similarly relies on geology and archaeology to make sense of the world; excavation is a primary means of obtaining information and reinventing the self. He joins with geologists and archaeologists to extol the power of the reader and writer to manipulate the traces of the past and thus to endure into the future.

Like most of his contemporaries, Dickens followed developments in geology and archaeology. His periodicals frequently published on both subjects,[1] which fact demonstrates that Dickens was undeniably familiar with contemporary discourse on the excavatory sciences. Demonstrating more than familiarity, he also took an active interest in geology and archaeology; he tackled both subjects directly in *Pictures from Italy*, published in 1846. He showed an interest in archaeology through his membership in the Antiquaries Club and his subscription to Charles Roach Smith's privately published *Illustrations of Roman London*.[2] Dickens was good friends with Austen Henry Layard, the most renowned archaeologist of the day, and also maintained a friendship with Sir Richard Owen, the celebrated paleontologist.[3] Dickens had a copy of Lyell's *Principles of Geology* in his library, and in *Darwin and the Novelists*, George Levine claims that geology was "part of Dickens's materials for imagining the world" (122). Like his fellow Victorians, Dickens struggled to make sense of a world newly understood to be vast and indifferent. In her essay on origins and oblivion, Gillian Beer writes, "Not only individuals but whole species take part in the immense and irrecoverable process of forgetting and being forgotten. Lyell's and Darwin's work raised the problem of how to sustain a narrative form that would satisfy the demand for coherence while acknowledging evanescence" (72). Dickens used geology and archaeology to reveal evanescence and to assert coherence.

"I Want to Know":
Seeking Foundations in *Little Dorrit*

Geological and archaeological relics are crowded into *Little Dorrit*'s London to illustrate time's stagnation in a novel in which the past haunts the present and impedes progress toward the future. Just as a Roman pavement incidentally discovered in the city impedes progress on urban improvements, Mrs. Clennam's pocket watch, a relic of her husband's youthful indiscretion, tethers her irrevocably to the past. Struggling through the temporal quagmire that they perceive around them, the characters in *Little Dorrit* find little movement or possibility of renewal: like Cavalletto imprisoned in Marseilles, time paces impatiently back and forth, unable to move on. J. Hillis Miller famously characterized *Little Dorrit* as a labyrinth: he wrote, "The image of the labyrinth suggests that life is not immobile enclosure but is endless wandering within a maze whose beginning, ending, or pattern cannot be perceived" (1965, 232). He could as easily have been describing the intellectual angst of the mid-Victorian, post-Lyellian period. Indeed, Dickens suggests that at the beginning of the nineteenth century time was dealt a crippling blow and, for many people and institutions, simply ceased to progress. Twenty years prior to the present of the novel, the Dorrit family entered the Marshalsea, and the Clennam family broke apart—Arthur's biological mother died, the connection between Arthur and Flora was severed, and Arthur joined his father abroad in China. At the same time, Maggy froze forever at ten years old, the novel's most poignant instance of temporal retardation. Dickens sets the restriction of time's movement at the point when Playfair's redaction of Hutton's *Theory of the Earth* popularly introduced uniformitarianism, and the paradigm of natural history and temporal progress irrevocably shifted. The key events that predate the action of the novel yet drive its plot coincide with the advent of geological time and introduce into *Little Dorrit* the oppression of anachronistic spaces.

Scientists and novelists alike complicated the power of a geological time with "no vestige of a beginning and no prospect of an end." They diminished its vastness and indifference with their emphasis on a sense of narrative guided by the uniformitarian theory of change. Dickens knows that the passage of a human life, even the passage of an entire civilization, represents only the smallest and most insignificant of movements on the vast time scale. Yet, like Tennyson, Dickens seems acutely aware of the paradox central to the uniformitarian view: while small actions are overwhelmed by the immense expanse of time, they evidently affect great change. Even as the individual withers, she discovers her power. The title, *Little Dorrit*, directs our attention to the little and indicates the potency of the small, of the "modest life of usefulness and happiness" (895) that Arthur and Amy finally achieve. In his essay on the novel, Mike Hollington discusses the slow movement that characterizes the novel. He notes the attention to detail and the painstaking descriptions of progress that several characters offer. He connects such attention to minutiae with the novel's preoccupation with size, observing that most of the characters who achieve goals are small—Pancks, Doyce, and Cavalletto. Of course, the title character ensures that we appreciate the authority of the small, the ability of the little to change the course of events. Hollington brings the moral superiority of these physically small characters and their accomplishments together with a description of time's incremental progress: "To do things 'inch by inch,' to take painstaking care over detail in the specification of time and change, is the corollary of finely tuned moral sensitivity, and of meticulous concern with real, other life. . . . Patient exactness is the key virtue in *Little Dorrit*" (118). Hollington's description evokes the uniformitarian sense of slow, incremental change over great expanses of time. We may anticipate that the characters in *Little Dorrit* will struggle to make a mark on time, but their incremental shifts will be vital and profound. Central to their movement is their relationship to the past.

The obscurity of beginnings and endings creates a desire for narrative, an "imposition of a plot on time" as Frank Kermode says (43).[4] In *Little Dorrit*, as in geology and archaeology, men and women excavate the debris of the past to uncover the fragmentary evidence that becomes the foundation for a plot. Dickens opens *Little Dorrit* with an image of time imprisoned and struggling to move forward. He sets the first chapter in a prison made suffocatingly close by the hot summer sun, and he personifies time's struggle against the bars of slow change. As Rigaud plans his escape from the Marseilles prison cell, Cavalletto paces back and forth across the cell, continually moving despite his confinement. Rigaud calls his cell mate a "clock" (42) and marvels that he always knows the time. Thus, the novel begins with a self-styled gentleman and a clock confined together in prison, and such a pairing proves a useful template for the themes that pervade the novel. People and time are entombed. Yet, Rigaud does escape and Cavalletto-time keeps moving. Later, the personification of time is struck down in a crowded London street, but, with Arthur's help, Cavalletto recovers from his injuries to move freely and effectively. Moving on despite the tendency toward stagnation and ruin is the great struggle depicted throughout *Little Dorrit*.

Bleeding Heart Yard is the site most sympathetically associated with this struggle. The yard is also the site most like the sites of urban reform with which Dickens's readers would have been familiar. Bleeding Heart Yard is a wreck, a decaying neighborhood, once great but now a shambles and the home of personal ruin and poverty. It is the sort of place becoming increasingly rare as urban improvements razed run-down buildings and neighborhoods, replacing them with thoroughfares and new construction. Bleeding Heart Yard appears to display decline: its antecedents, remaining at this site only in the obscurity of legend, are far grander than its contemporary state. An old royal hunting ground dating from the days of William Shakespeare, Bleeding Heart Yard is "a place much changed in feature and fortune, yet with some relish of ancient greatness about it" (176). The "maze of shabby streets" characterizes both the contemporary slum and the physical manifestation of years of urban reform. Bleeding Heart Yard is associated with the old, run-down parts of London that were rapidly being replaced by the railway and other innovations. Writing of Holywell Street, not far from the actual location of Bleeding Heart Yard,[5] Lynda Nead echoes Hillis Miller: "It obeys the logic of the labyrinth; it is illegible and multifarious, as opposed to homogenous and purposeful" (163). Even the name of the neighborhood illustrates its descent from former stature: once called "Bleeding Hart" to reflect the success of the royal hunts, the name devolves to describe the impotence of the bleeding hearts who wither there, unable to pay their rent or move on to higher ground. An outstanding example of a Ricouerian trace, Bleeding Heart Yard reveals at once the strata that show years of urban change and is itself a product of that change.

Reflecting the stagnation that permeates Dickens's London, the physical layout of Bleeding Heart Yard suggests the inability of the working-class inhabitants to improve their social welfare. Bleeding Heart Yard's sunken position makes progress nearly impossible for those who live there. The yard maintains its original level, while the rest of the city rises up around it. As urban layer replaces decaying urban layer, the yard lies at the bottom of a stratified metropolitan space. By remaining static, the stature of this neighborhood sinks, while the city around it aspires to higher levels: "As if the aspiring city had become puffed up in the very ground on which it stood, the ground had so risen about Bleeding Heart Yard that you got into it down a flight of steps which formed no part of the original approach, and got out of it by a low gateway into a maze of shabby streets, which went about and about, tortuously ascending to the level again" (176). Dickens offers a geological image of this segment of the urban landscape: like a cliff face that reveals layers of rock, each representative of a distinct epoch, this cluttered area of London reveals through its layers a glimpse of the city's history. This physical embodiment of the past existing, if not thriving, still in the present appealed to Dickens, who often expresses a nostalgia for what Michelle Allen has called the urban picturesque. Allen describes the resistance to the Thames embankment and sanitary reform, noting the "aesthetic appreciation of the unreformed, degraded river re-imagined as a site of the picturesque" (89). Similarly, Bleeding Heart Yard is the sort of neighborhood characterized by the "variety and irregularity of the structural forms" (Allen 90) traditionally associated with the picturesque and here transplanted by Dickens into an urban context. Dickens clearly values the urban picturesque: the yard's pastoral past gives it a strong foundation, and its architectural multifariousness is a celebration of the diversity of urban life.

Indeed, Bleeding Heart Yard is ultimately a site of progress rather than decay. The yard "was inhabited by poor people, who set up their rest among its faded glories, as Arabs of the desert who pitch their tents among the fallen stones of the Pyramids" (Dickens 1967, 176).[6] These characters, notably the Plornishes, are among the best examples in *Little Dorrit* of characters grounded in the past yet slowly moving on toward a brighter future. When their fortunes change for the better, thanks to the generosity of William Dorrit, Mrs. Plornish refurbishes the corrupted ruins of the yard into a neo-Shakespearean bungalow: "[painted to represent the exterior of a thatched cottage] . . . it presented . . . a little fiction in which Mrs. Plornish unspeakably rejoiced" (630). This renovation incorporates the past, reinterpreting it not as a ruin but rather as the root of modern industry. The Plornishes reinvent the ruins that surround them. In fact, although this neighborhood embodies what Nancy Metz calls an "architecture of decay" (1990, 465), Dickens does offer the poor inhabitants a model of progress. As one ascends the stairway to exit the yard, one must pass Doyce and Clennam's factory, a symbol of productivity and improvement.[7] Positioned above the exit, the factory intimates that ascent comes from the real progress of "metal upon metal." In Bleeding Heart Yard, Dickens offers a familiar site of urban ruin, a site he explicitly associates with the accidental archaeology so prevalent in mid-century London, but he recasts that site as a foundation for progress. Dickens offers this small, picturesque trace within London as a model of reform despite circumstances and in contrast to so many other spaces in the novel.

Time in Bleeding Heart Yard is marked by the accumulation of material: in the Clennam and Casby homes, time seems to have stalled altogether. The oppressive ticking of time that will not pass resonates audibly in the Casby house, where Dickens paints in miniature the stasis that permeates Victorian London—against the backdrop of geological time, human time is inconsequential, while time in general looms and taunts. Arthur is overwhelmed by time's crushing presence in the Casby house, itself an enormous clock, every part of it ticking persistently: "There was a grave clock, ticking somewhere up the staircase; and there was a songless bird in the same direction, pecking at his cage, as if he were ticking too. The parlor-fire ticked in the grate. There was only one person on the parlour-hearth, and the loud watch in his pocket ticked audibly" (186). The caged bird is a powerful symbol of what it is to inhabit the Casby house, and, more, what it is to inhabit the London of *Little Dorrit*. The bird cannot escape, and his nature is so stifled by his prison that he cannot sing but can only mark the passage of time by relentlessly beating out the minutes on the bars of his cage. Petrified by his house-watch, like a fossil or an archaeological artifact, Casby has been "unchanged in twenty years," and has been "little touched by the influence of the varying seasons" (186). A portrait of the patriarch as a young boy reveals that Casby remains nearly identical to his former self: "There was the same smooth face and forehead, the same calm blue eye, the same placid air" (187). Of course, the older Casby's hair has turned gray and he has a bald spot absent in the pictured boy, but these developments seem cosmetic rather than substantive. In essence, the boy and the man are the same; they even coexist on the parlour-hearth where both Casby-past and Casby-present greet visitors, who inevitably note the patriarch's substantial immunity to time's signature. For Casby, time is less a prison than a preservative. Even Arthur refuses to note the passage of the years in Casby's face: when Casby admits, "We are older, Mr. Clennam," Arthur responds, "We are—not younger" (188). Casby and his house encourage Arthur to negate time's forward trajectory.

While Casby seems indifferent to time's glacial pace, his daughter manipulates her language to assert control over time. Flora concocts absurd speech patterns, flitting from past to present, crashing through temporal boundaries at every turn. When Arthur returns from China, Flora reclaims the old affectionate terms of address the two shared in their youth and, at the same time, acknowledges how much their relationship has changed. In the following typically muddled passage, she moves indiscriminately from her past to her present lives:

> Meanwhile Flora was murmuring in rapid snatches for his ear that there *was* a time and that the past was a yawning gulf however and that a golden chain no longer bound him and that she revered the memory of the late Mr. F and that she should be at home tomorrow at half-past one and that the decrees of Fate were beyond recall and that she considered nothing so improbable as that he ever walked on the north-west side of Gray's-Inn Gardens at exactly four o-clock in the afternoon. (Dickens's emphasis; 201)

The syntactic confusion here reflects Flora's efforts to move freely through time. Her emphatic past tense in the first sentence—"*was*"—asserts the distance of the past, but her sentiment attempts to revive that past. She also interjects quite precise references to possible meeting times in the immediate present—"half-past one" and "exactly four o'clock"—amidst recollections of her separation from Arthur and also her marriage to the late Mr. F. The last sentence quoted above illustrates how crafty Flora has become in her manipulations of time: her use of "ever walked" implies the past tense, but she actually suggests an appointment in the future. Her reference to such a precise moment in time confirms what we begin to suspect: Flora strategically applies tense shifting and garbled syntax to achieve a modicum of control over time. Flora voices the confused sense of different times existing at once: the "yawning gulf" and "half-past one" describe the geological time and the everyday time that Victorians struggled to reconcile.

In the Casby house the oppressive ticking reminds the inhabitants of their persistent struggle against time. In the Clennam house, it appears the struggle has ended: Mrs. Clennam is entombed in the past. She utterly refuses to move into the present; as she explains, "the track we have kept is not the track of time; and we have been left far behind" (85). Arthur attempts to act as an archaeologist as he endeavors to excavate his mother's stony silence, but he is met with unyielding rock. Upon returning to his mother's house after a long absence, Arthur observes that nothing there has changed; in fact, the decoration in the house suggests that the somber constancy has endured much longer than the twenty years of Arthur's absence: "The old articles of furniture were in their old places; the Plagues of Egypt, much the dimmer for the fly and smoke plagues of London, were framed and glazed upon the walls" (72). Victorian London linked thus with ancient Egypt, and both crowded into the tomb-like Clennam house, reminds Arthur and the reader that the past weighs heavily indeed on this family. Dickens even casts Mrs. Clennam herself as a relic from ancient Egypt: when Arthur questions her about the past, she relapses into stone (86) and demonstrates "the impenetrability of an old Egyptian sculpture" (87), her frown so fixed it is "as if the sculptor of Old Egypt had indented it in the hard granite face, to frown for ages" (87). Demonstrating her remove from the present and the forward movement of time, Mrs. Clennam asks Arthur if it snows; she boasts, "I know nothing of summer and winter, shut up here. The Lord has been pleased to put me beyond all that" (74). Mrs. Clennam exists beyond the passing of the seasons. She embodies stasis, her crippled body representing crippled time, unable or unwilling to move on.[8]

Dickens emphasizes Mrs. Clennam's destructive relationship with time when he makes central the pocket watch that haunts her bedside. With its insistent message—"D[o] N[ot] F[orget]"—it reveals the root of Mrs. Clennam's emotional imprisonment:

> No, sir, I do not forget. To lead a life as monotonous as mine has been during many years, is not the way to forget. To lead a life of self-correction is not the way to forget. To be sensible of having . . . offences to expiate and peace to make, does not justify the desire to forget. Therefore I have long dismissed it, and I neither forget nor wish to forget. (406)

She never uses the words "remember," "remind," or "recall," which would suggest an activity abhorrent to Mrs. Clennam—re-membering literally means piecing together again, an occupation familiar to archaeologists. Instead, her life, as she describes it, is devoted to not-forgetting, to dwelling in the past and refusing to move into the present.

When Mrs. Clennam finally reveals the truth the watch insists she not forget, Rigaud impatiently demands she move on to the part of the story about money, a much more practical and quantifiable obligation than the debts of sin which long governed Mrs. Clennam's life. He cries out, "Come, madame! Time runs out. Come, lady of piety, it must be! You can tell nothing I don't know. Come to the money stolen, or I will! Death of my soul, I have had enough of your other jargon. Come straight to the stolen money!" (846–47). Four times in this short passage he commands that she arrive at a part of the story, the only part, he claims, that has any relevance to the present; he insists that she "come" to the present and to the fiscal matter of debt owed to the Dorrit family and the blackmail he intends to collect to conceal that debt. With his brief and to the point, "Time runs out," Rigaud disparages Mrs. Clennam's long commitment to the past. But she is not bullied by his threats or his imperative emphasis on the present. She retains control of her story, and this is key. As does Flora, Mrs. Clennam clearly asserts control over time, in her case the past, through narrative. Her resistance to Rigaud's blackmail has little to do with money and much to do with her refusal to allow him to tell her story. When Rigaud starts to tell the story long concealed by Mrs. Clennam, she violently intercedes, ". . . with a bursting from every rent feature of the smouldering fire so long pent up" (842). Strikingly, Dickens uses a catastrophic metaphor: the sudden eruption of a volcano supercedes the long, slow process of gradualism—one geological metaphor replaces another. She cries out, "I will tell it myself! I will not hear it from your lips, and with the taint of your wickedness upon it. Since it must be seen, I will have it seen by the light I stood in. Not another word. Hear me!" (842–43). Flintwinch protests, but she insists: "I tell you, Flintwinch, I will speak. I tell you, when it has come to this, I will tell it with my own lips, and will express myself throughout it" (843). The repetition of the first person pronoun and of the words, "tell," "speak," and "express" combine to present a strong speaking subject. In this role, Mrs. Clennam reclaims her past from Rigaud's corruption as well as from the ravages of time. Near the conclusion of her tale, she explicitly acknowledges that she interprets the relic-letters "D.N.F." according to her own ends: "He died, and sent this watch back to me, with its Do not forget. I do NOT forget, though I do not read it as he did. I read, in it, that I was appointed to do these things. I have so read these three letters since I have had them lying on this table, and I did so read them, with equal distinctness, when they were thousands of miles away" (846). She interprets the fragmentary remains of the past—a trace—as she wishes, and thus she proves her authority over the past: she interprets the story and tells it in her own way.

Mrs. Clennam's climactic report of past events satisfies the reader who has wondered at the mysterious relationship between the Clennam family and Little Dorrit, but Arthur never learns the truth his mother explains to Rigaud. The book opens with his return home and with his expressed aim of learning to interpret the story told by the pocket watch. In this opening, Dickens clearly casts Arthur as an archaeologist, a figure who seeks the past through the alchemy of excavation and subsequent storytelling. He likens Arthur to Giovanni Battista Belzoni, an explorer who had imported many impressive Egyptian artifacts to London in the early part of the nineteenth century; Belzoni's discoveries were displayed in the Egyptian Hall in Piccadilly in 1821, roughly contemporary with Arthur's return to England.[9] Dickens suggests that his protagonist also excavates ancient tombs and unravels mummies when he writes that the vacant churches Arthur passes in the City "seemed to be waiting for some adventurous Belzoni" (70) to dig them out and interpret their significance. Although Belzoni's Egyptian finds were displayed thirty years before Dickens wrote *Little Dorrit*, he brought those ruins back to the surface of public discourse with the publication in 1851 of "The Story of Giovanni Belzoni" in *Household Words*.[10]

The figure most popularly associated with the ruins of ancient civilizations was Austen Henry Layard. In fact, for Dickens, Layard represented one
of *Little Dorrit*'s peculiar and powerful conjunctions: the relationship between
archaeology and administrative reform. The most famous archaeologist of
the mid-nineteenth century, Layard took London by storm with his *Nineveh
and Its Remains* (1849), considered the first archaeological best seller.[11]
Layard and Dickens were close personal friends and traveled together in Italy
in 1853.[12] In 1855, when Dickens was writing *Little Dorrit*, he joined Layard,
then an MP, in a campaign to reform administrative inefficiency. Inspired by
the horrific and fatal administrative blunders of the Crimean War, Layard's
association was part of a larger movement that began in 1854 with a blue
book on administrative reform written by Charles Trevelyan and Stafford
Northcote.[13] Among other reforms, Trevelyan and Northcote advocated
what would have amounted to a civil service exam, the idea being that administrative work should be clarified and professionalized. Dickens satirizes
the horrible disorder of civil bureaucracies run like gentlemen's clubs and
staffed largely by indolent younger sons with the Circumlocution Office and
the Barnacle family. He makes explicit the source for these elements of *Little
Dorrit* in the preface to the 1857 edition, in which he refers to his involve-
ment in the Administrative Reform Association: "If I might offer any apology
for so exaggerated a fiction as the Barnacles and the Circumlocution Office,
I would seek it in the common experience of an Englishman . . . in the days
of a Russian war . . ." (35).[14]

The Circumlocution Office, thus, is not only a morass of administrative
inefficiency, but is also one of the novel's central archaeological sites. Dickens
states boldly, "Numbers of people were lost in the Circumlocution Office"
(146); after passing through other government departments over a "slow lapse
of time and agony," they arrived at the Circumlocution Office and "never reap-
peared in the light of day" (147). It is a labyrinth, a tomb that only the most
skilled archaeologist can penetrate, navigate, and escape. Asserting the cultural
distance of a true archaeologist—"Allow me to observe that I have been for
some years in China, am quite a stranger at home" (152)—Arthur strives to
extract information from this adamantine relic of a bureaucracy; he aims to excavate this site, meticulously sifting through layer after layer of paper debris,
but he must first chip away at the encrusted strata of Barnacles that impede access to the truth. It's worth mentioning that Charles Darwin published a two-
part monograph on barnacles in 1851 and 1854, just as Dickens began work on
Little Dorrit, and the choice of the name Barnacle seems to be deliberate.[15]
After impressively filing down the resistance of the lesser Barnacles, Arthur
manages to see Mr. Tite Barnacle in his home. Barnacle is unable or unwilling
to provide any substantive information. Instead, he directs Arthur back to the
government office: "The Circumlocution Department, sir, . . . may have possibly recommended—possibly—I cannot say—that some public claim against
the insolvent estate or firm or co-partnership to which this person may have

belonged, should be enforced" (153). Barnacle's insistent qualification of each nonspecific statement demonstrates a larger principle of circumlocution: at every turn in *Little Dorrit*, information is obscured and difficult or impossible to recover. Yet, Arthur devotes himself to the recovery of information, and he is presented as an archaeologist in his attempts. Rodney Stenning Edgecombe agrees: "Indeed the whole treatment of the Circumlocution Office resembles the interpretive, reconstructive procedure of an archaeologist bent on clarifying by tentative guess as much as by solid inference a civilization altogether mysterious and remote" (69).[16]

To excavate administrative layers, one requires access to records, and Arthur finally chips away enough Barnacles to catch a glimpse of the fossilizing records buried in their care. He badgers the young Barnacle into forwarding him to the proper office with his persistent demand, "I want to know" (154). He penetrates deeper and deeper into the office, pounding his way through with his refrain—"I want to know"—until he finally meets a surprisingly useful person who directs him toward the true work of the social archaeologist, paperwork. And here his excavation comes to a frustrating end. The troubling oxymoron of paperwork characterizes the stasis that cripples the Dorrits, the Circumlocution Office, Mrs. Clennam, and so many others in Dickens's London: paper does not work. Administrative offices cannot relieve debt or provide patents—they cannot even report accurate histories—and novels cannot resolve. The archaeologist who studies only document-relics will find nothing. Despondent and with no further digging to do, Arthur leaves the Circumlocution Office only to run into Mr. Meagles. These men began their acquaintance while traveling abroad, specifically while visiting Egyptian tombs, and now they meet again under circumstances which Dickens would have us see are not at all so very different.

The Clennam and Dorrit stories, rather like *Little Dorrit* itself, frustrate Arthur and Dickens's reader alike; they share an absence of beginnings and endings and offer very little to those who seek to uncover their truths. Arthur returns to England set on uncovering the origins of his family's business trouble, but instead he devotes himself to the Dorrits. Yet, however much he tries, Arthur ultimately fails to learn the history of the Dorrit debt, and he never knows what Amy and the reader learn about his own origins. An experienced reader of novels expects Arthur to uncover the identity of Dorrit's creditor— perhaps even Arthur's mother—and somehow resolve the conflict, but Dickens breaks with expectations. Sylvia Manning and Janet Larsen both address Dickens's narrative strategies and his refusal to provide closure. Larsen describes characters' repeated attempts to end things and the persistent postponement of the end; she mentions specifically the revolution invoked with references to Marseilles but never brought near fruition within the novel (42). She offers Dickens's own model of circumlocution as a description of the novel's movement: the novel dwells in a "deadly waiting-time with no vision of origin and end that would give meaning to the suffering" (44). Manning also considers the notion of stasis: "the text presents the possibility of a world in which all we are really doing is circumlocuting" (127). Both critics claim that Dickens deliberately defies novelistic conventions in *Little Dorrit*, the form of the novel mirroring its major theme of evasive resolution. Manning describes the novel's end:

> The confusion of the denouement enforces an impression that all the threads are being pulled together, but if we pause to unravel the detail we see that the convergence is not there. The conventions of plot, that we very much expect Dickens to uphold . . . require signification in the juxtaposition of events. At the end of *Little Dorrit* the story breaks loose from this ideology. (136)

In *Little Dorrit*, origins are obscure and resolution remains out of reach. Janice Carlisle explains: "The relative uncertainty embodied in this ending is appropriate. The characters in *Little Dorrit* are confronted with a world that is incomprehensible" (211). Carlisle is absolutely right to connect the lack of resolution in the novel to the incomprehensibility of the world, but Manning and Larsen too easily reduce the relationship of that world to time: Arthur and Amy strive to escape stasis, to penetrate circumlocution, and to comprehend. Like the Megalosaurus that waddles up Holborn Hill, the past intrudes into the present and obscures notions of progress. Dinosaurs and engraved pocket watches will conquer the present unless excavators recover their meanings and retell their stories.

With its refusal to uncover origins and its failure to secure resolution, *Little Dorrit* participates in a uniformitarian model of time that may frustrate archaeologists but would nonetheless strike them as very familiar. Geological time moves in imperceptible increments, with "no vestige of a beginning and no prospect of an end." Manning inadvertently borrows the language of geology but fails to appreciate the significance of the temporal model she describes: "The novel has foiled our expectations for mystery as for adventure, expectations derived from our indoctrination in the ideology of nineteenth-century novelistic plot, from our belief in beginnings and endings" (141). What Manning neglects to consider is that many of Dickens's readers had already been divested from their belief in beginnings and endings by the work of geology and archaeology. Writing in 1857, Fitzjames Stephens (1829–1894) echoes the language of uniformitarianism in his review: "the artistic fault of *Little Dorrit* is that it is not a tale. It neither begins nor ends—it has no central interest, no legitimate catastrophe, and no modeling of the plot into a whole" (qtd. in Manning 143). Dickens's failure to begin or end reflects the new understanding of the Earth and it defies expectations of narratives. Arthur's archaeological failures reflect the limits of excavation as a means of understanding, but Amy's productive use of ruins offers an alternate and more hopeful possibility.

The ancient Egyptian culture of tombs and stony statues represented by Mrs. Clennam and excavated by Arthur-Belzoni stands in marked contrast to the culture of ancient Rome, which also figures prominently in *Little Dorrit*. In Dickens's view, Egypt represents the insurmountably foreign, so far away, spatially and temporally, that one can never recover a perfect understanding of its remains. Dickens suggests a coevalness between Victorian London and ancient Rome, but he discovers no such relation between ancient Egypt and contemporary life. His allusions to Egyptian statuary evoke only museum displays, not any familiar pattern of life: Mrs. Clennam's stony visage never cracks; Arthur never learns her secret. Amy is the one who uncovers the truth Mrs. Clennam has devoted herself to not-forgetting. In a confessional letter, Arthur's mother reveals her late husband's infidelity and her own role in taking her husband's child from its mother and raising it as her own.[17] The letter represents Mrs. Clennam's final attempt to impose a plot: she narrates her life and Arthur's in an effort to escape Rigaud's blackmail. She finally not only tells her story, she writes it. In authoring the narrative she has long concealed and preserved, she assumes an authority Rigaud cannot defy.

But Amy can. Amy is the audience of Mrs. Clennam's letter, and, like an archaeologist, Amy has the authority to interpret this trace (that is, the woman and her confession both). Instead of broadcasting the letter's contents, and thus increasing her own wealth and probably rescuing Arthur from the psychological burden of an unloving "mother," Amy chooses to suppress its narrative. Mrs. Clennam explicitly authorizes Amy to reveal the contents of the letter:

> The great petition that I make to you . . . the great supplication that I address to your merciful and gentle heart, is, that you will not disclose this to Arthur until I am dead. If you think, when you have had time for consideration, that it can do him any good to know it while I am yet alive, then tell him. But, you will not think it; and in such case, will you promise me to spare me until I am dead? (858)

Though she asks Amy to wait to tell Arthur what she has learned, the very fact that she poses the question gives Amy the authority to answer and to decide herself what is best. Mrs. Clennam goes on to explain why she prefers Arthur to remain ignorant: "Let me never feel, while I am still alive, that I die before his face, and utterly perish away from him, like one consumed by lightning and swallowed by an earthquake" (860). She uses an image from catastrophist geology, and Amy answers her with uniformitarianism. Though she honors her future mother-in-law's request, she evidently feels that some artifacts are better buried. Amy does not kill Mrs. Clennam, but she destroys the evidence of her narrative. Amy obliterates Mrs. Clennam's past when she asks Arthur to burn the letter of confession. In so doing, she ensures that Egyptian history remains opaque. In contrast, Rome is knowable: in its ruins, nineteenth-century visitors recognize the antecedents of their own culture.

When Dickens first describes Rome in *Little Dorrit*, he seems to emphasize how the decay of the city recalls the ruin and particularly the stasis that pervade London. In his discussion of the Dorrits' arrival in Rome, Angus Easson observes how pointedly Dickens presents the site as one of decay: "we reach a place full of history, or rather, one whose history has apparently ceased, except in the apocalyptic prospect of destruction. As the Dorrits approach Rome, the landscape degenerates, spatially rather than temporally . . . and in Rome ruin and stagnation are the only principles" (37). The Roman ruins signify a failure to progress, a failure also demonstrated by just about every character in the novel. Indeed, Rome appears to be "a city where everything seemed to be trying to stand still for ever on the ruins of something else" (Dickens 566). The landscape the Dorrits survey offers nothing; it is "a repetition of the former Italian scenes, growing more dirty and more haggard as they went on" (566). While Dorrit embraces the prospect of progress with the decision to go on a tour, the repetition of the landscape undermines the very notion of a voyage as a progressive journey. The journey across the Alps and into Italy disappoints: it leads nowhere. The peculiar conjunction of the living travelers with the frozen dead at Great Saint Bernard is a striking illustration, reminiscent of Pompeii's plaster casts: frozen, still, and utterly unsalvageable, these travelers rest at the lodge unaware that they will never move on. Instead of leading to an array of new perspectives and new prospects, the Dorrits' Italian tour is an exercise in entropy.[18] Traveling wears them down to a state of ruin no different from the decay that surrounds them: Fanny compromises her happiness in her marriage to the unremarkable Sparkler, Tip wastes himself in gambling, and finally, Edward and Frederick Dorrit die.

Amy Dorrit is the only character who returns from Rome undiminished. By eliding past and present, she successfully builds a life for herself on ruined foundations. In her article on "Usable Pasts," Alison Booth argues that Amy personifies possibility:

> Without these personified possibilities [Amy and Dorothea Brooke], the history narrated in these novels would take the form of wave upon wave of dashed ambition. The Clennams and Lydgates of the world would face perpetual defeat, as households, enterprises, and discoveries collapse in material and moral exhaustion. The heroines, the potential of mercy and love, must stand clear of the rubble, disregarding—in ignorance or courage—the manifest lessons of history, so that all may read a tolerable, intelligible narrative of progress. (205)

Amy does not keep clear of the rubble, rather she delves into its solitude, but she reads the Roman ruins as signs of her own life and times.

In this respect, Amy interprets ruins not as remains of the past but as potential foundations for the future, much like the Plornish family in Bleeding

Heart Yard. It is significant that Amy reflects on the Roman ruins at such length in the two letters she writes Arthur from Rome. These letters predate Mrs. Clennam's confession, but they function similarly: they represent the author's attempt to control her life and make order out of an emotionally chaotic situation. Amy is overwhelmed by her changed fortune and her family's newfound liberty; she is deeply disappointed in her father's behavior; and, on top of these difficult adjustments, she is beginning to recognize her feelings for Arthur and feel sadness at his interest in Minnie Gowan. Her two letters are divided fairly equally between delicately concerned descriptions of Minnie's welfare, designed to keep her own emotions in check, reminiscences of her old life, and travel-narrative style representations of her experiences in Rome. The latter are the most grounded, though these, too, function primarily to make sense of Amy's new world, a world she more comfortably associates with old ruin than with new fashion. She writes incredulously to Arthur of her own ignorance: "Old as these cities are, their age itself is hardly so curious, to my reflections, as that they should have been in their places all through those days when I did not even know of the existence of more than two or three of them, and when I scarcely knew of anything outside our old walls" (609). Amy expresses greater interest in the ruins that existed during her own lifetime than in the old life of Rome. She even collapses the distance between the ancient empire and her own life in London when she applies the same modifier, simply "old," to both.

Amy does not hesitate to associate herself with the ruins, even taking comfort amongst the fragments. Edgecombe writes, "Little Dorrit, having been stripped of her past by Fanny and by her father, uses antiquarian materials to bring back that deleted past, and so to nourish and sustain the soul" (72). Left alone by her father who returns to London for business and by Fanny lately married, Amy takes solace in the ruins, which she visits on her own. These visits, in fact, seem to be the only visits made to such sites by any of the Dorrits; the rest of the family shows no interest in ruined Rome, and perhaps such indifference reflects their ignorance of their own variable states of ruin. In contrast, Amy recognizes the similarity of her present and past landscapes:

> Little Dorrit would often ride out in a hired carriage that was left them, and alight alone and wander among the ruins of old Rome. The ruins of the vast old Amphitheatre, of the old Temples, of the old commemorative Arches, of the old trodden highways, of the old tombs, besides being what they were, to her were ruins of the old Marshalsea—ruins of her own old life—ruins of the faces and forms that of old peopled it—ruins of its loves, hopes, cares, and joys. Two ruined spheres of action and suffering were before the solitary girl often sitting on some broken fragment; and in the lonely places, under the blue sky, she saw them both together. (671)

The connection to the Marshalsea follows the mention of "old tombs," a reference that evokes the tomb and prison imagery that pervades the novel. Yet, to Amy, the tomb and the prison are home; they offer the comfort of the familiar. Imprisoned, Amy could help her family; in Rome, she is superfluous. While the rest of the Dorrits live in elaborate denial of their poverty and limitation, Amy builds a small but meaningful life under the shadow of the Marshalsea. She finds solace in the Roman ruins, for their stony stasis evokes her old life, whereas the fine parlors and "prunes and prisms" of her new life remain insurmountably foreign.

Amy mourns the loss of her former life, a life imprisoned amidst people fallen victim to financial and social ruin; and she sees, in the Roman fragments scattered about her, that familiar life itself ruined. In these fragments, she sees greatness fallen to pieces: she and her family indeed live in ruin despite their financial solvency and their luxurious accommodations. The ruins she finds in Rome remind her of her happier past and of the sudden dissolution of her past, but also of the fragility and falsity of her present. Fanny's marriage, Mrs. General's protocols, even her father's pleasure in his new wealth all prove to be shallow and subject to one form of ruin or another. Of course, Amy does not foresee such social collapse when she sits among the Roman fragments, but she does perceive the state of moral ruin in modern society, a society that makes her long for the comforting ruin of her impoverished days. Away from the daily struggle of keeping her father comfortable and her brother and sister morally and profitably employed, Amy is exposed to a vast abyss in which no single life can possibly matter. This is the reality so many of Dickens's readers faced in a world irrevocably expanded by geology. At perhaps the most celebrated archaeological site of all—Rome itself—Amy cannot turn away from the depths that excavation reveals. However, she can interpret the remains and thus evade oblivion. Gillian Beer writes of the restorative power of interpretation in the face of oblivion:

> . . . it may be that we should see the Victorian insistence on the 'real' as in some measure a response to the loss of a close-knit beginning and ending in the natural world. No longer held in by the Mosaic time order, that history became mosaic of another sort, a piecing together of subsets into an interpretable picture. Words like *traces* and *decipherment* become central to geology, evolutionary theory, and fictional narrative at this time. Interpretation was the only assurance. (1986, 68)

She can read Roman ruins as the detritus of her own life, thus conflating her past and an empire's. In so doing, she collapses time and aggrandizes herself. Rather than being paralyzed by the awesome specter of an almost infinitely deep past, Amy instinctively knows how to use the past, how to conquer its vastness and make it her own. Her ability to build a productive life in the Marshalsea and her keen sense of the significance of the Roman ruins set Amy apart from the other characters in *Little Dorrit*. Rather than being corrupted by ruin, Amy builds upon fragments and provides one of the only examples of progress in the novel. She asserts agency over her own narrative despite the burden of time: to comfort her confused and dying father, Amy readily reverts to the past when she assumes her former role and even her old costume; later, she as easily moves forward when she guides Arthur to incinerate the only record of his origins.

Dickens seems to contrast Amy's use of the past with Meagles's brazen, albeit comic, exploitation of artifacts; fusing them into his contemporary English life, Meagles almost completely disregards their origins. He collects objects from all the famous sites, notably Egypt and Pompeii, but his souvenirs reflect the present and his own culture more than they reveal anything about the past. Meagles's "miscellany" (236) consists of specious relics from ancient civilizations:[19]

> There were antiquities from Central Italy, made by the best modern houses in that department of industry; bits of mummy from Egypt (and perhaps Birmingham); . . . morsels of tesselated pavement from Herculaneum and Pompeii, like petrified minced veal; ashes out of tombs, and lava out of Vesuvius; . . . rosaries blest all round by the Pope himself, and an infinite variety of lumber. (236–37)

That Meagles's mummies may have been made in Birmingham[20] demonstrates the character's gullibility and complacent worldly ignorance. Writing of Layard's frustration with the "less discerning side to the burgeoning wealthy middle-class cultivation of the arts" (1986, 182), Michael Cotsell describes the prevalence of forgeries and the imposition on eager but undereducated collectors; he locates Meagles in this class of naifs and argues that Dickens "associates the assumption of connoisseurship with mental slavishness" (184). But Meagles's collection was never meant to signify the past. It reveals less about his artistic judgment than it does about the value Meagles places on his unshakable ability to remain wholly English when abroad. He prides himself on his inability to speak French and other languages, and his collection reflects the spirit of his travel: he conquers foreign sites and distant times and imports their relics into his home. Indeed, the conquest of the past is so complete that the traces of the past can be fabricated in the present. Furthermore, the ruin of the present is as much of interest to Meagles, and to Dickens, as is the ruin of the past, if not more so. Indeed, for all his misuse of the past, it is Meagles who finally facilitates the exchange of information that reveals the connection between the Clennams and the Dorrits. Perhaps Meagles's collection marks him as an experienced excavator, one who can bring traces of the past home and assimilate them into a progressive, modern life.

Meagles's collecting and Amy's reflecting on the parallels between Roman ruins and her own life are part of Dickens's insistent conflation of ruined Rome with Victorian London. Just a few years after the Great Exhibition, when life in England seemed to embody the very spirit of progress, the parallel with ruins would not necessarily have been a common view,[21] yet the connection between Rome and London pervades every description of Rome in the novel. Cotsell observes that the English in Italy are always of greater interest than the Italians: "It is a noteworthy fact that Dickens does not demonstrate a great interest in Roman ruins in *Pictures from Italy*, whereas they are very much a part of his re-flections in *Little Dorrit*" (195). When Italy is his subject, Dickens is not espe-cially interested in ruins, but when England is his subject, those ruins take on a striking significance: "the subject is always not just Italy, but the English in Italy, and by implication, or in memory, the English at home" (195). In essence, Cotsell argues that the ruins in Italy reflect the ruin of English society. Indeed, Dickens's characters frequently recast Rome in English terms: the English in Rome continue to live by "English time" (Dickens 652), and Meagles justifies his trip to visit Pet with a variation on a cliché—"Just as Home is Home, though it's never so Homely, why you see . . . Rome is Rome, though it's never so Romely" (581). Meagles links English home and ruined Rome with rhyme, and the logical extension of the proverb is that Rome is actually not much like Rome, but, is, in fact, more like home.

Indeed, as Nancy Metz points out, Dickens's description of Rome in *Pictures from Italy* overtly likens the ruined city to Victorian London. She argues that the comic analogy Dickens offers in his travel narrative foreshad-ows the serious comparison he draws in *Little Dorrit* (1990, 479). Metz quotes a telling selection from *Pictures from Italy*:

> When we were fairly off again, we began, in a perfect fever, to strain our eyes for Rome; and when, after another mile or two, the Eternal City appeared, at length, in the distance; it looked like—I am half afraid to write the word—LONDON!! There it lay, under a thick cloud, with innumerable towers, and steeples, and roofs of houses, rising up into the sky, and high above them all, one Dome, I swear, that keenly as I felt the seeming absurdity of the comparison, it was so like London, at that distance, that if you could have shown it me, in a glass, I should have taken it for nothing else. (qtd. in Metz 479)

Interestingly, the Rome Dickens presents here, as Cotsell noted, is not the Rome of ruins but the living Rome of the present. Dickens evokes Rome's preservation and its ability to weather time when he uses the appellation "Eter-nal City." He thus emphasizes the city's longevity rather than its ruin, and he associates such longevity with London when he insists upon the visual similar-ity between the two cities. He goes farther to make the comparison personal: when a Londoner, such as himself, looks in a glass, he sees Rome. In other words, the likeness between Rome and London transcends the appearance of the two and has serious implications for the inhabitants of the English city.

Learning to live in time and to rely on the power of the little is finally the only way to break free from stasis. Arthur's archaeological excavations end when Amy burns the only record of his true past, and when his speculative ventures fail, Amy secures his release from debtor's prison. From the beginning of the novel, she is the only character who knows how to make her way through ruin, whether literal or financial. She shows only passing interest in the distant past, and no great concern over the future; she strives only to progress gently through the present—the parish records that mark her past and presumably her future are, to her, only pillows to provide comfort on a long night. Once married to her, Arthur also adopts an incremental, gradual approach to time, and the two of them, in an ending many readers misconstrue as unhappy, move gradually through the ruin of London, affecting small reforms: "They went quietly down into the roaring streets, inseparable and blessed; and as they passed along in sunshine and shade, the noisy and the eager, and the arrogant and the froward and the vain, fretted and chafed, and made their usual uproar" (895). These final words of *Little Dorrit* describe an acceptance of uniform time, an acknowledgment of the individual's very small yet very significant impact in time.[22]

Interestingly, the concluding paragraph of *Little Dorrit* echoes a key passage from Darwin's *Voyage of the Beagle*. Darwin describes with awe his experience of observing uniformity in the rocks of a streambed:

> Amidst the din of rushing waters, the noise from the stones, as they rattled one over another, was most distinctly audible even from a distance. This rattling noise, night and day, may be heard along the whole course of the torrent. The sound spoke eloquently to the geologist; the thousands and thousands of stones, which, striking against each other, made the one dull uniform sound, were all hurrying in one direction. It was like thinking on time, where the minute that now glides past is irrecoverable. So was it with these stones; the ocean is their eternity, and each note of that wild music told of one more step towards their destiny. (303)

The small steps toward destiny are told in small stories and simple deeds. The acceptance that the past is irrecoverable and that progress toward the future is dull and uniform is *Little Dorrit*'s happy ending.

Excavating *Our Mutual Friend*: "The Power of Knowledge"

The digging in *Our Mutual Friend* is driven not only by the desire to know the past, as is also the archaeological work in *Little Dorrit*, but also by a desire to eliminate ruin and thus embrace present and future. In *Little Dorrit*, Arthur struggles to understand his father's past actions; in contrast, rather than focus on the waste of Old Harmon's life, *Our Mutual Friend* emphasizes his legacy, his son who focuses far more on building his own family than on refiguring his father's past. Old Harmon is never explained or redeemed, but Baby Bella becomes the centerpiece of the novel's end and the hope for the future. The actual legacy, that is the final and legally binding will, is rewritten in the present: the remains of the Harmon estate are disseminated at the will of those who survive, and the actual will of the deceased is removed from the ruin only to be obliterated by the good will of the Boffins. In other words, the central characters in this novel rebuild, reconstruct, and ultimately evade ruin in a way the characters in *Little Dorrit* did not.

Many readers of *Our Mutual Friend* have considered the metaphoric significance of the dustheaps, a central site in the novel and the material inheritance of first the Boffins and later John Rokesmith (a.k.a. Julius Handford and John Harmon).[23] The dust has been read as a general indicator of change, associated with both Darwin and urban reform; it has been considered a representation of corrupt wealth and social climbing; and it has most generally been seen to represent the waste of life at mid-century.[24] One could argue that the dust means any one of these things, or that it means many things at once; alternately, one could argue that the symbolic resonance of the dustheaps shifts awkwardly throughout the novel. For instance, the dust represents the taint of wealth, associated with the Veneerings, the Lammles, and Fledgby, but also the honest industry of Boffin and Rokesmith, even Venus. In short, with the dustheaps Dickens crafts a signifier that is a perfect cipher—at once a powerful symbol and a complete nothing. In other words, the dust means so much that it becomes identified less with any specific meaning and more with the construction of meaning, of story, of knowledge.

I offer a new reading of *Our Mutual Friend* with my surprisingly unusual claim that while they may represent much more, the dustheaps are also simply dustheaps. I will show that the substance of the dust is less important than the action it invites, indeed, requires—that is, excavation. As Howard Fulweiler succinctly explains: "Both capitalism and natural selection are imaged in the mounds, partially composed of fossils that are at the same time saleable garbage—filthy lucre" (66). As he explains, the dustheaps contain multiple metaphoric meanings, and while it may at times be useful to extract and examine only one of those meanings, we must remember that the dust in its very essence is multifarious, literally and metaphorically. Just as real and fictional dustmen and women sort the dust to find an array of materials, readers of *Our Mutual Friend* must sort through the dust to find an array of meanings. The value of dust lies not in its ability to contain one particular sort of item but in its ability to contain so many items. The reader or excavator who sorts the dust gleans a greater understanding if she keeps in mind the layers and the multitude of objects and meanings contained within the dust. Dust may represent waste in a larger, metaphoric sense, and it surely represents "filthy lucre" as well, but above all it is simply dust and is thus something that must be excavated and ultimately cleared away.

In addition to the simple experience of living in the city and thus being familiar with the fact of dustheaps, Dickens may have had particular encounters with these urban phenomena in mind when he planned the novel. Henry Mayhew (1812–1887) offers a sociological view of the dust industry in *London Labour and the London Poor* (1861–1862), which extended report surely informed Dickens's knowledge and opinion of the subject. This objective account was partnered in Dickens's experience with two personal contacts with dust contractors and their yards: one located near Regent's Canal and the other in Nova Scotia Gardens. In "Dust; or Ugliness Redeemed" (1850) published in *Household Words*, R. H. Horne (1803–1884) describes a particular dust yard, near what is now King's Cross Station, peopled by three lively characters who introduce the reader to the trade. The yard runs alongside a canal, and is thus certainly Henry Dodd's yard which was in the neighborhood of Regent's Canal, also the location of the Harmon dust yard. K. J. Fielding writes: "It has also been said, on fairly good authority, that the character of Boffin was at least partly suggested by a celebrated dust contractor, Henry Dodd [1801–1881], whose business was conducted at Hoxton" (118). It is interesting to note that the "fairly good authority" Fielding cites is none other than Charles Roach Smith. In his *Retrospections: Social and Archaeological* (1883), Roach Smith includes a short chapter on Henry Dodd, and it seems fitting that this "benevolen[t]" dust contractor (Roach Smith 98) offers the most direct connection between Dickens and Roach Smith.[25] According to Michael Cotsell in *The Companion to* Our Mutual Friend, Dickens became acquainted with Dodd in 1858 when Dodd attempted to make a charitable gift of five acres of land to the Royal General Theatrical Fund: Cotsell and Fielding suppose that Dodd's generosity found its way into *Our Mutual Friend* in the form of Boffin's generous spirit (Cotsell 57). Dickens had another encounter with a dust contractor around the same time: through his consultations with Angela Burdett-Coutts (1814–1906) regarding her charitable work, Dickens became familiar with the dust yard located in a slum called Nova Scotia Gardens, which Burdett-Coutts purchased and eventually transformed into a series of model housing units.[26]

Mayhew offers the most comprehensive survey of dustmen and their trade. He enumerates quantities and types of dust and goes on to discuss the character of the men, women, and children who deal in dust. Unlike some of the London poor he interviewed and described, the dust workers receive no special sympathy from Mayhew. Like Dickens, he seems more impressed by the dust itself than by those who cart and sift it. He expresses awe at what the dust could do were it left to accumulate. In his first paragraph of "Of the Dustmen of London," Mayhew claims that without dust collection all of London would be buried beneath debris: "Now the ashes and cinders arising from this enormous consumption of coal would, it is evident, if allowed to lie scattered about in such a place as London, render, ere long, not only the back streets, but even the important thoroughfares, filthy and impassable" (218). Thus, the dustmen protect London from the progress of time: decayed cities and extinct species cannot avoid the gradual accumulation of debris that eventually buries them, leaving them accessible only to the geologist or archaeologist. Yet, the inhabitants of London escape this fate as long as the debris of their daily lives is removed to designated grounds, leaving the landscapes of everyday cleared and free from the waste that indicates the inevitable passage of time.[27] Mayhew goes on to calculate that "the gross amount of 'dust' annually produced in London would be 900,000 tons or about three tons per house per annum" (226). If this amount of dust were not carted away, the city would soon resemble Pompeii, buried in ash, albeit ash industriously and systematically, rather than catastrophically, produced. Mayhew makes explicit this comparison: "the dust is piled up to a great height in a conical heap, and having much the appearance of a volcanic mountain" (229). In likening a dust heap to a volcano, Mayhew employs a rhetorical strategy common in the discourse that surrounded dust: as I discuss below, Horne and Dickens both employ the same strategy. The simile illustrates how awesome these dust mounds were—they are volcanoes, ancient, angry geological formations suggestive of catastrophic explosion and prehistoric violence more familiar to dinosaurs than to men. They thus reveal how closely connected modern life is to antediluvian times. Recall how naturally that Megalosaurus waddled up Holborn Hill. Dust heaps are that lizard's natural habitat. Yet, at the same time, there is something empowering in presuming that humans can generate, and generate as a mere by-product of daily life, something as terrible and impressive as a volcano. Mayhew's representation of dust, though undeniably empirical, establishes a pattern of using dust to connect the urban landscape to the past and to the sciences of excavation.

In "Dust; or Ugliness Redeemed," Horne also describes the dustheap as a geological landscape: "It was in fact a large hill, and being in the vicinity of small suburb cottages, it rose above them like a great black mountain. Thistles, groundsel, and rank grass grew in knots on small parts which had remained for a long time undisturbed" (380). The dustheap is not only likened to a mountain but to an old mountain, one that has been a feature of the urban landscape for a "long time." When Dickens introduces his own dustheap nearly fifteen years later, he presents it as a geological site. Early in *Our Mutual Friend*, Mortimer Lightwood describes the Harmon dustheaps: "On his own small estate the growling old vagabond threw up his own mountain range, like an old volcano, and its geological formation was Dust" (24). The language here is explicit: Dickens links the landscape of dust with the geological landscapes brought so vividly before the Victorian public in the mid-century. Yet, the Harmon dustheaps are decidedly distinct from true geological formations. The creator of this mountain range, this old volcano, is neither the uniform process of slow change over countless millennia nor is it God. The old vagabond Harmon accumulated the dust and heaped it on his estate. Thus, Dickens offers a far more complex rendering of the faux artifacts in Meagles's collection; here we have an entire geological formation that is man made. The role of men in making, surviving, and making sense of ruin is very much the subject of *Our Mutual Friend*.

Horne's dustheap has a grandeur and permanence capable of sustaining life. Not only can plant life grow on the mound, but the debris that comprises the mound can be turned to a purpose. Joel J. Bratten argues that Dickens was likely influenced by Horne's emphasis on redemption. In his enumeration of the various materials to be found in the dustheaps, Horne repeatedly stresses the recycled use of those materials. For example, "'the bones' are selected with care, and sold to the soap-boiler. He boils out the fat and marrow first, for special use, and the bones are then crushed and sold for manure. Of 'rags' the woollen rags are bagged and sent off for hop-manure; the white linen rags are washed, and sold to make paper, &c." (380). Like Mayhew, Horne catalogs the ingredients of the dustheap and explains how each material is used.[28] Fine cinders, for example, are sold to make bricks; vegetable and animal matter are sold for manure; oyster shells are sold to make new roads (380). In this early form of recycling lies redemption: recall Horne's subtitle, "Ugliness Redeemed." Even waste is not wasted in an industrious society, and thus this feature of the urban landscape signifies not filth and decay but rather industry and progress: "everyone is hard at work near the base of the great Dust-heap" (350).

The hard work and the dust itself combine in Horne's account to bring a man back from the dead. Horne recounts how three dust workers spy a body in the canal: they pull the man from the water and bury him up to the neck in the ashes of the heap, the warmth of which revives him. When he regains consciousness, he discovers that a piece of parchment just unearthed has the power to restore to him a portion of his lost fortune. Thereafter, he becomes acquainted with the dust contractor and his daughter to whom he is eventually married, increasing his fortune by the 20,000 pounds given her by her father. Thus, the dust thrice redeems the life of this man, first literally giving him life, then a living, and finally a fortune and a wife. In the three characters who facilitate this man's redemption we see a clear source for Dickens's *Our Mutual Friend*: Horne's one-legged dust sifter, though a woman, prefigures Silas Wegg; the bone collector named Gaffer Doubleyear is likely the namesake of that more gruesome scavenger Gaffer Hexam; and "little Jem Clinker, a poor deformed lad whose back had been broken when a child" (Horne 379) is a boy version of the similarly deformed and sympathetic Jenny Wren. Of course, the man who is at the center of the narrative may well have inspired Dickens's protagonist: like John Rokesmith, this man is nearly drowned and through the kindness and industry of dust workers eventually comes into his fortune and

secures his beloved wife. While the similarity of characters in Horne's jour-
nalistic piece and Dickens's novel is remarkable, more interesting is the similar emphasis on the power of the dust itself. In both cases, the nearly drowned
man is saved not only by the kindness of others but also by the material prop-
erties of the dust.

Another possible source for the dustheap in *Our Mutual Friend* is the former dust yard cleared to make way for the housing project, Columbia Square.
This site embodies redemption in replacing the filth and decay of the yard
with the cleanliness and progress associated with the new building. The writer
of "Hail Columbia-Square," published in *All the Year Round* in 1862, describes
the area of Bethnal Green once known as Nova Scotia Gardens: "The air is
perverted to carry from window to window the monstrous vapours encircled
in a compound interest of pollution as it passes on. The sun's rays are perverted, and instead of bringing wholesomeness and purity with them, draw up
new wealth of nastiness from every nook and corner, and heating it to fever
pitch, breed Death far and near" (302). To emphasize the horror of living
amidst such squalor, the author begins with a series of striking contrasts—
a man "may get up rich and lie down poor . . . at six o'clock in the forenoon he
may be a father, and before the clock has made its round he may be childless"
(301)—and then adds to the list the contrast between the "rows and rows
of palaces" in South Kensington and the "narrow" and "tortuous" streets and
lanes of Bethnal Green. He describes at length the filth associated with the
rooms of the poor, so foul that they "taint" the very air and sunlight that grace
the homes on the other side of town: "This chamber . . . may be supposed to
vomit out its impurities" (302). The implication is clear: the sunny, open
homes of South Kensington may be fouled by the contagion vomited forth by
Bethnal Green. The writer then turns to the "goodly work" of Angela Burdett-Coutts, who the reader will perceive not only restores dignity to the poor
but also rescues the wealthy from the taint of such filth.

Burdett-Coutts financed and oversaw the design of Columbia Square,
a housing project built atop a razed dustheap.[29] The praise lavished on the
tenements by the writer in *All the Year Round* emphasizes the cleanliness of the
dwellings: "It must be remembered that the very object of building these
houses was to bring together an enormous number of very poor people, never
accustomed to live where any amount of cleanliness was so much as possible"
(304). In contrast, the Columbia Square buildings admit quantities of light
and air that exceed those enjoyed by residents of Bloomsbury, or even May-
fair (304), and a laundry that runs the full length of the attic makes cleanliness
accessible to a class of citizens who has never known such a luxury. The
language in the article makes explicit the contrast between the new model
tenement and the old dustheap that once lay in its place: "Accumulation of
dirt . . . must be carefully guarded against, and light and air must be everywhere" (304). Indeed, the residents of Columbia Square are physically, even
mechanically, removed from the accumulation of dirt. The writer describes

the waste disposal facility: "A trap opens in the floor of each corridor, down which the inhabitants of the different rooms shoot their dust and refuse, into a great dust-hole underneath. Surely, every one who lives in a house where the dust-bin is so placed as that the dustmen have to carry its contents through the house to get to their carts, has to envy the inhabitants of this Columbia-Square" (304). The residents are thus removed from their waste by the trap and however many floors of the building and are even better situated vis-à-vis their dust than their wealthier counterparts. Clearly, the Burdett-Coutts reformation of Nova Scotia Gardens into Columbia Square not only provided the poor with improved accommodation but also removed them from the dust that so contaminated the region before. I present the *All the Year Round* discussion of Columbia Square in such detail because it demonstrates the exemplary reform Dickens may have had in mind when he fictionalized the removal of the Harmon dustheaps. Clearing debris offers the hope of discovering valuables, but it is also a valuable activity in its own right, clearing the ground for improvement. Boffin's ironic name for his home amidst the dust heaps—the Bower—is stripped of its irony once the dust is removed, and this good man is left to live happily in a place that now deserves its name.

For Mayhew and Horne, dust yards signify redemption, and we see their influence on Dickens. Yet, I argue that the attention to redemption—dust is not waste but is actually a productive and profitable substance—distracts from the material presence of the dust mounds. While the improvement of the land that formerly housed the dustheaps is implicit in the novel's resolution,[30] Dickens more emphatically attends to the work of sifting, or excavating, the dust and thus gaining knowledge about the past and about identity in the present. Dickens was primarily interested in the entirety of the mounds and in the labor associated with them. As Fielding points out, "The skillful organization of 'collecting dirt' on a large scale is, in fact, 'a kind of work' that deserves a very high reward—much higher than share-pushing, money-lending, and shady finance" (119). The kind of work is a labor akin to archaeological and geological excavation and carries the weight of those scientific labors, labors that reveal the past and the value of the present at once. Howard Fulweiler speaks to the distinction I seek to make:

> Like the geological record, the mounds are composed of old bones as well as other disparate objects whose position and value in the mounds have no plan. Instead, they simply accumulate. All of society contributes to the pile, and the dustheap accumulates slowly by chance and random selection. . . .
> The digging in the mounds by Silas Wegg and others leads naturally to an analogy in paleontology. The digging is a search through the fossil record of Victorian London. It is here that a central concern of the Victorian novel comes together with the chief goal of Victorian science: uncovering the secret of inheritance. (61–62)

Fulweiler makes a compelling argument about inheritance, but I believe that "digging," "search[ing]," and "uncovering" are the key words of this passage. Far more important than the dust itself or even the valuables buried therein is the act of excavation.

Harmon and the other characters who sift the dust, Boffin and Wegg most prominently, function as archaeologists or geologists. In his discussion of the novel's debt to Darwin, Fulweiler notes the publication in *All the Year Round* of a review of Lyell's *The Geological Evidences of the Antiquity of Man*:

> The reviewer focuses on the evidence from Danish islands: "heaps of waste oyster-shells, cockle-shells, and waste of other edible shell-fish, mixed with bones of divers eatable beasts and birds and fishes. . . . The Danes call these mounds—which are from three to ten feet high, and some of them a thousand feet long by two hundred wide— kitchen middens." These "refuse heaps," mixed with instruments and fragments of pottery, are an interesting analogue to the dustheap, the central symbol of *Our Mutual Friend*. (54)

Fulweiler makes the point that Dickens's view of urban life is heavily influenced by Darwin's theory of evolution, particularly the notions of accumulation and mutuality. I would add that Dickens's use of the dustheaps as a central image in the novel reveals his interest in the potency of geological sites. In other words, Lyell describes the inclusion of prehistoric humans amongst geological remains, and Dickens, likewise, considers the dustheaps—piles of remains, of traces from which the living might profit—the foundation and the context for human life. Fulweiler argues that the mounds, like biology and geology, offer evidence that reveals inheritance: "It is in the solution to the question of inheritance, considered not only in its Darwinian biological aspect but in a larger, human way, that *Our Mutual Friend* will arrive at its ultimate meaning" (62). Fulweiler argues that Dickens connects biological and financial inheritance through the dustheaps; he is most interested in what is learned from the dust. I find his conclusions compelling, but I am more interested in *how* Dickens's characters learn from the dust. Different characters learn different things, but they all learn through the process of excavation.

Though Fulweiler focuses primarily on geology and evolutionary biology, the passage he quotes from *All the Year Round* certainly describes the sort of site that would be attractive to archaeologists, who would examine the remains of human life found there. Dickens extends the point: he, along with many of his generation, imagines not only human possessions but humans themselves as contributing to the content of the dustheaps. Horne experiments with this notion when his dust workers revive the nearly drowned man by burying him up to the neck in the dust. When his condition is improved, he is dug out again—excavated and returned to life, a personification of Pompeii's plaster casts, given life through the narrative imagination of archaeologists and writers. Dickens's association of people with dust generally produces a narrative much bleaker than Horne's redemptive one, though a nearly drowned man is also central in Dickens's account. When Mortimer Lightwood and Eugene Wrayburn set off for the Hexam residence to investigate the apparent drowning of the Harmon heir, their carriage rolled, "down by where accumulated scum of humanity seemed to be washed from higher grounds, like so much moral sewage" (30). The accumulated scum is associated at once with the waste of dustheaps (recall the controversy over whether or not these heaps contained excrement) and with geological waste, washed away in the natural process of erosion. The geological association is in many ways the more damning one, for it invokes not just the waste of poverty but the inevitable waste of all human life. Dickens continues in this vein when Mortimer and Eugene enter the Hexam house, which "had a look of decomposition" (31). The people who live along the river, according to this description, live amongst the rotting decay of modern urban life and are themselves rotting away. Dickens thus associates them with both sewage (and dust) and with a scientific understanding of nature which is itself rife with decay. Dickens makes explicit the connection between natural and human decay with Noddy Boffin's selected reading: Gibbon's *Decline and Fall of the Roman Empire* (1776). Bringing this text into his reader's imagination invites the reader to compare the dust exposed in London with the ruin of the Roman Empire, a comparison that preoccupied many archaeologists and popular writers, as discussed in chapter 4.

Dickens broadens the scope of the connection between humanity and decomposing remains beyond the riverfront with its poverty. Charley Hexam, for instance, is raised amidst the "moral sewage" and does not seem to have escaped its taint. Even through the kind efforts of his sister, Lizzie, Charley remains a corrupt and cruel young man. In his chapter "Salvage and Salvation," Efraim Sicher considers Charley as an example that illustrates how savagery and civilization are reversed in *Our Mutual Friend*. Sicher notes that Dickens describes Charley thus: "a curious mixture in the boy, of uncompleted savagery, and uncompleted civilization" (Dickens 28). As Charley grows more educated and improves his socioeconomic prospects, he is physically removed from the literal and metaphoric waste of his home neighborhood, yet he seems to grow less civilized and more savage in his temperament.[31]

Certainly, in coming under the influence of Bradley Headstone, Charley finds himself in the company of one who is disconnected from the filth of extreme poverty yet is fatally connected to the murderous impulse of "nature red in tooth and claw." Headstone fights ferociously to condemn Eugene's corpse to the river's waste: had Lizzie not applied her scavenging skills to rescue Eugene,[32] his body would have rotted and decomposed in the Thames and eventually would have been discovered along with broken crockery, oyster shells, and animal bones. It is not ironic but rather just that such decomposition should be the ultimate fate of Headstone himself, alongside the no less villainous Rogue Riderhood. The two in their mortal struggle fall into a "smooth pit" and are found at some time in the distant future, "lying under the ooze and scum behind one of the rotting gates" (781). They become nothing more than geological remains.

Yet, not all the characters in *Our Mutual Friend* are condemned to rot in pits, subjects for excavation. Indeed, those that live their lives fully, happily, and productively are those that articulate their relationships to the past and their identities in the present through excavation. As discussed above, Lizzie Hexam achieves her Cinderella happy ending by excavating Eugene's body from the river, and of course the wills that provide the Boffins and the younger Harmons with their happy endings are excavated from the dustheaps. Along with these acts of excavation—that is, extracting something from the earth—Dickens attends to the work of the excavator who must make sense of the fragments he has discovered. This sense-making is central to the novel and is most often manifested as articulation. In "The Artistic Reclamation of Waste in *Our Mutual Friend*," Nancy Aycock Metz writes that the novel "is less concerned with the resolutions toward which social issues are tending than with the multiple and continuous acts of putting the world together that the individual imagination performs" (60–61). As I say above, the process of sifting the dust is more important to Dickens than what happens to the dust or even than what is found within the mounds. Metz describes complementary methods of making sense: "Articulation and analysis are ways in which the imagination seeks to formulate and answer the question, what is the meaning of this?—the one by piecing fragments together, subordinating the parts to an imagined whole, the other by stripping down superfluities to reach what is essential" (67). Piecing fragments together is a central act for geologists, archaeologists, and writers of fiction. Barthes would have us see that Michelet also emphasizes articulation as a form of sense-making and, in his view, history-making: he describes the "historian-as-bone-collector and restorer of human dust" (84). Dickens makes the piecing together of fragments a central act for his characters in *Our Mutual Friend*.[33]

The most obvious example of one who pieces together is Mr. Venus, who makes a living as an articulator and is quite literally a bone collector.[34] Venus describes himself as an anatomist, but Wegg invites him to moonlight as an archaeologist, recognizing the vocational link between excavation and articulation. He "expatiates on Mr. Venus's patient habits and delicate manipulation; on his skill in piecing little things together; on his knowledge of various tissues and textures; on the likelihood of small indications leading him on to the discovery of great concealments" (300). This intellectual conjunction of the shard of evidence and the narrative it implies reflects the definition of the archaeologist as put forth in 1851 by Charles Newton (1816–1894) at the Oxford meeting of the Archaeological Institute: "The plodding drudgery which gathers together his materials, must not blunt the critical acuteness required for their classification and interpretation" (26). Modeled on a St. Giles's taxi-dermist's shop shown to Dickens by his illustrator Marcus Stone (1840–1921), Venus's shop contains a peculiar potpourri of animals stuffed and set in poses (such as a pair of dueling frogs poised forever in conflict in the shop window) and human bones and body parts, babies in jars, and skeletons. When Wegg first approaches the shop "in a narrow and a dirty street," he observes "a mud-dle of objects vaguely resembling pieces of leather and dry stick, but among which nothing is resolvable into anything distinct" (83). Of course, Wegg's point of view is certainly not one associated with searing clarity; however, Dickens is clear that when we first meet Venus, his shop is yet another in-stance of the accumulated waste of life in London. It presents a collection of fragments not yet articulated in any meaningful way.

Lurking in the corner, the French gentleman, a skeleton reconstructed out of assorted bones collected from many sources, serves as a useful barome-ter. In this introduction to Venus and his shop, the skeleton is fragmented. When Wegg inquires after his own leg, which Venus evidently acquired when it was amputated at hospital, Venus "takes from a corner by his chair, the bones of a leg and foot, beautifully pure, and put together with exquisite neat-ness. These he compares with Mr. Wegg's leg; that gentleman looking on, as if he were being measured for a riding-boot." Wegg recoils from this foreign limb, and Venus explains that it belongs to "'that French gentleman,' whom [Wegg] at length descries to be represented (in a very workmanlike manner) by his ribs only, standing on a shelf in another corner, like a piece of armor or a pair of stays" (85). The skeleton is incomplete, seeming to consist only of a rib cage, a leg, and a foot. Yet, what few bones Venus has acquired, he has as-sembled "beautifully" and with "exquisite neatness." This bodes well. Indeed, as the novel progresses, the skeleton grows more and more complete, its frag-mentation eventually giving way to a useful wholeness.

I would even suggest that as Venus turns away from the path of blackmail and destruction, the French gentleman becomes a more solid construction.[35] He transforms from a collection of bones to a haunting figure with a "grin upon [his] ghastly countenance" capable of "computing how many thousand slanderers and traitors array themselves against the fortunate" (570). He adds his own gruesome support to Venus's eager move away from bringing ruin to another, and he even participates in the preference for science over "wiggery" (565). One reason Venus offers when he extracts himself from Wegg's corrupt project, is that he prefers to return to his scientific work: "'I am taken from among my trophies of anatomy, am called upon to exchange my human wari-ous for mere coal-ashes warious, and nothing comes of it'" (472). When Wegg suggests that the vastness of the mounds offers encouragement, Venus re-sponds: "'They're too big. . . . What's a scratch here and a scrape there, a poke in this place and a dig in the other, to them?'" (472). He correctly perceives the position of humans as small and insignificant when set against the vast backdrop of the geological landscape, and he prefers to assemble fragments—like his now grinning French gentleman—rather than bury himself in the infinite task of uncovering remains.[36]

Like Princess Ida, Venus turns away from the vastness to tend to the domestic and to marriage, the pleasant assembly of just two parts into a loving whole. Just before Venus, Harmon, and Boffin entrap the scheming Wegg, Wegg and Venus talk in the shop, and when Wegg turns to look into the corner for the French gentleman, he notices that the shop has been cleaned. It is no longer a site where nothing resolves; through the feminine touch of Venus's fiancée Pleasant Riderhood, it has become tidy and resolute, as has Venus, who has not only become moral but has also had a haircut. Venus obliquely tells Wegg about his pending marriage, and the dialogue reveals that Venus now extends articulation into the verbal realm. When Wegg inadvertently insults Venus by referring to Pleasant as "the old party," Venus rejects the slang: "in a case of so much delicacy, I must trouble you to say what you mean" (762). His shift to this alternate form of articulation is explicit in his earlier conversation with Boffin: ". . . having prepared your mind in the rough, I will articulate the details" (565). We should anticipate Venus's turn to the good when we first meet him and he returns Wegg's amputated leg, commenting, "I am glad to restore it to the source from whence it—flowed" (295). He expresses a preference for reconstruction, for making whole what had been fragmented. As his name suggests, Venus values union above disaggregation: parts are precious only if they contribute toward a whole.

Like Venus, Dickens, too, relies on narrative to articulate, to assemble fragments. In his discussion of Dickens's use of physics, Jonathan Smith argues that Dickens offers fiction as a way to resist the dissipation threatened by the second law of thermodynamics.[37] He, the writer, need not subject himself to the scientific patterns at work in the world around him, whether those be entropic forces or gradual change; he can assert meaning and impose moral order at will. Smith concludes:

> I would argue that reliance on, and the calling attention to, the "conventions of narrative" represents in *Our Mutual Friend* a partially conscious expression of uneasiness at the secretive intervention necessary to force an unruly second-law universe back into a comfortable, optimistic, ordered first-law model. . . . In retreating to the comparative safety of the first law in *Our Mutual Friend*, Dickens covertly acknowledges that this order is a fiction, a work of articulation that cannot put off death but can make some use of the bones. (1995, 66–67)

Smith sees the most explicit connection between articulation and Dickens's novel in its serial publication: the chapters, numbers, and books are patched together to form a whole.

Like Venus turning to narrative with Boffin and eventually to correcting the particulars of Wegg's verbal articulations, and Jenny Wren who eventually sets aside her piecework dolls' dress-making to tell the recovering Eugene restoring stories, Dickens celebrates the value of narrative as the most lasting way to hold together fragments. He dramatizes this most empathically in the story of John Rokesmith who strives to reconstruct a life where there seemed to be only ruin. Rokesmith assembles fragments most explicitly when he presents himself to Boffin as a secretary. Because Boffin is unfamiliar with such a position, he invites Rokesmith to demonstrate the skills: "if you'll turn to at these present papers, and see what you can make of 'em, I shall know better what I can make of you" (181). Rokesmith proceeds to sort the papers into organized heaps—the choice of the word "heap" seems designed to suggest a connection between this paper-work and the excavation associated with the dustheaps:

> Relinquishing his hat and gloves, Mr. Rokesmith sat down quietly at the table, arranged the open papers into an orderly heap, cast his eyes over each in succession, folded it, docketed it on the outside, laid it in a second heap, and, when that second heap was complete and the first gone, took from his pocket a piece of string and tied it together with a remarkably dexterous hand at a running curve and loop. (181)

The description of Rokesmith's secretarial work is important because it forges a connection between the three motifs I have been discussing: excavation, articulation, and sense-making. As my own reader will know, these are the primary phases of archaeological work. Excavation leads to the discovery of artifacts, often fragments, which are then pieced together, or rearticulated into a cogent whole; finally, someone reads the object and reports its story. In the brief passage describing Rokesmith's organization of Boffin's papers, Rokesmith engages in each one of these phases, although not in such a neat chronological fashion. In sifting through papers, carefully labeling items, and moving from heap to heap, he works as an archaeologist: recall the representation in *Little Dorrit* of paperwork in the Circumlocution Office as archaeological. In sorting the papers into sensible piles, he pieces together fragments, and in producing a logically organized stack of documents complete with abstracts, he makes sense of the fragments.

This brief example of Rokesmith's interview with Boffin offers in miniature a pattern that is associated with this character throughout the novel: he seeks and then organizes information, and he crafts narratives, some fictional and others not. Of course, the most obvious fiction he creates is that of his identity. He presents himself as Julius Handford when he goes to view the supposed corpse of John Harmon and as a nameless "strange man" when he goes in disguise to interrogate Riderhood. In both instances he conceals his identity in order to gain information. In fact, even before events turned dire with the murder of the man dressed in Harmon's clothes, he had concealed his identity to gain information about his bride-to-be. Thus, such a clandestine method of gaining knowledge was not precipitated by the desperate circumstances of the murder but rather seems to Rokesmith/Harmon to be a legitimate means of intelligence gathering. Even when his identity is revealed to the Boffins and to the reader, the ruse continues and is made more complex by the masquerade of Boffin's miserliness all for the continued purpose of getting a good read of Bella's character. Essentially, Rokesmith finds meaning in his life, not through direct means but by creating fictions in order to make sense of his problematic situation.[38] His name is a small but wonderful example: a "smith," of course, is a craftsman, and "roke" is a mist or vapor; thus John is a self-made crafter of mist. He makes a name that signifies the difficulty of signification.

In Rokesmith's keynote chapter, "A Solo and a Duet," he attempts to piece together what he knows about the events that led to his supposed death. Dickens forges a clear connection between the assorted disguises this man has assumed and his ability to make sense of what has happened when he prefaces Rokesmith's account with a striking physical description:

> Here he ceased to be the oakum-headed, oakum whiskered man on whom Miss Pleasant Riderhood had looked, and, allowing for his being still wrapped in a nautical overcoat, became as like that same lost wanted Mr. Julius Handford, as never man was like another in this world. In the breast of the coat he stowed the bristling hair and whisker, in a moment. . . . Yet in that same moment he was the Secretary also, Mr. Boffin's Secretary. For John Rokesmith, too, was as like that same lost wanted Mr. Julius Handford as never man was like another in this world. (359)

In the next sentence, he connects the fourth identity with the others: "I have no clue to the scene of my death." The point is made clear two paragraphs later when he speaks imperatively to himself: "Don't evade it, John Harmon; don't evade it; think it out!" (360). The reader now knows that Harmon, Rokesmith, the strange man, and Handford are one and the same. The assortment of identities makes Rokesmith himself a site for excavation and also enables him to excavate more effectively his own story. A significant portion of the chapter is a first-person, chronological narrative recounting Harmon's return to England. The many interjections—"Now, stop, and so far think it out, John Harmon. Is that so? That is exactly so" (360)—reveal that this narrative is a work in progress. Its completion will ultimately reveal or definitively conceal John Harmon.

The question Rokesmith poses to himself—"Should John Harmon come to life?" (366)—suggests that he may choose to excavate himself, that is, to dig through the accumulated layers of identities and pull to the surface the true, original self.[39] The reason to do so seems to have little to do with the murder or the inheritance and a great deal to do with Bella. The chapter continues as a duet between John and Bella, and her part leaves him certain that as a pair these two are out of tune, and Harmon must stay buried. Rokesmith presents himself as a geological relic, one he hopes to keep concealed:

> He went down to his room, and buried John Harmon many additional fathoms deep. He took his hat, and walked out, and, as he went to Holloway or anywhere else—not at all minding where—heaped mounds upon mounds of earth over John Harmon's grave. His walking did not bring him home until the dawn of day. And so busy had he been all night, piling and piling weights upon weights of earth above John Harmon's grave, that by that time John Harmon lay buried under a whole Alpine range; and still the Sexton Rokesmith accumulated mountains over him, lightening his labour with the dirge, "Cover him, crush him, keep him down." (372)

Dickens could hardly forge a clearer connection between the identity of an individual and an object of excavation; yet, unlike most objects, this one has a will and prefers not to be unearthed. Dickens thus offers an interesting twist on the anxious response to geological evidence: instead of dreading the loss of the individual to the inexorable power of geological forces, Harmon annihilates himself.

But, of course, he is finally revealed to his wife Bella and to Mortimer Lightwood.[40] The true story is told only when Rokesmith has arranged events in the present so that they are free from the taint of the past. He only allows the trace of his former self to emerge when he is confident that he has rewritten himself. He and Bella live contentedly and already have produced a new Harmon, thus moving on toward the future rather than looking back to the past. Once domestic happiness is secured, the products of excavation—John Harmon himself, his father's will, Boffin's true character—can be brought to light and can be neatly arranged within the narrative that the mutual friend has crafted.

In *Little Dorrit* and *Our Mutual Friend*, Dickens explores the two alternatives offered by urban archaeology: Londoners could make life amongst the ruin, finding comfort in the crumbling walls of their present as they fall away to reveal the remains of an imperial past, or they could, like Venus and Rokesmith, find what they required in the ruin and then use those fragments to move on. Amy Dorrit walks easily amidst the Roman ruins; John Harmon waits until the ruins have been removed or reconstructed to begin his new life. The people of nineteenth-century London vacillated in their approach: destruction and construction appeared synonymous, and excavation was the science that allowed the delicate equilibrium of life lived in the present, not cowed by the past nor alarmed by the future—quiet, content, and alive.

6

Final Fragments

I began with the trace, which has echoed through this work and has modulated as traces do. The trace is a relic from the past, it is an object in the present, it is a marker of time's passage, it is the point where time and space come together, it is the paradoxical evidence of the subjectivity of objects as interpretations shift and relics mean differently over time. Traces are fossils and bones, shards of pottery and partial mosaics; traces are also people and pasts and presents and stories. Above all, as Ricoeur says, the trace endures, and this has been a study of endurance. The Victorians faced one of the greatest paradigm shifts we have on record: the bottom dropped out of time, and they had to reinvent their relationship to the Earth, to time and history, and to God. This new world view was nearly shattering and produced some of the most lovely writing I know—"The individual withers and the world is more and more"—as men and women of the nineteenth century struggled to matter. *Excavating Victorians* assembles evidence from geology, archaeology, and the writings of Tennyson and Dickens to assert that the Victorians succeeded: they were not shattered. They endured.

In her essay "Origins and Oblivion in Victorian Narrative," Gillian Beer observes that when they finally turned away from the Mosaic chronology offered by the Bible, the Victorians took comfort in a different sort of mosaic: they faced the world through "a piecing together of subsets into an interpretable picture" (68). They assembled fragments and articulated them, materially and through narrative. It is the conjunction of trace and interpretation that saved the Victorians from deep time and nature red in tooth and claw. The observer in Lyell's margins, Macaulay's New Zealander, Tennyson's frame-poet, Amy Dorrit, Mr. Venus, Jenny Wren, and John Rokesmith are all excavating Victorians. They regard the ruins, sift the dust, articulate the pieces; and what they craft is not only a sense of the past but, much more importantly, a sense of themselves. They tell stories, and through narrative they rescue the individual from oblivion.

Retracing my own footsteps and ending where I began, I will turn back to Matthew Arnold and "Dover Beach." Like Tennyson and Dickens, Arnold stands small yet bravely on that darkling plain and offers the miniscule but powerful force of love to combat the grating roar of pebbles and their eternal note of sadness. His "Ah, love, let us be true to one another" seems feeble, yet it resonates, perhaps only because the reader so wants it to, but it resonates nonetheless. The roar closes up time like a fan, as Hardy put it, bringing Arnold's speaker and Sophocles together: both found "in the sound a thought" and this excavation of the sound, this discovery of something therein, is what protects the individual from annihilation. The inexorable passage of time, and so much more time than any mind can fathom, roars deafeningly in the Victorian ear, yet Arnold's speaker maintains his interpretive presence and finds a thought. What the thought is remains vague in "Dover Beach," but it does not matter. All that matters is that in finding the thought, asserting an interpretation, the individual raises his voice against the roar. The trace and interpretation are brought together through excavation, and such excavation gave the individual Victorian voice.

Notes

1. Introduction:
"All Relics Here Together"

1. Hutton expanded his theory in his book *Theory of the Earth with Proofs and Illustrations* (1795). This work, in turn, was translated into clearer prose by John Playfair in his *Illustrations of the Huttonian Theory of the Earth* (1802). Playfair's was the most widely read of the three publications. In *Time's Arrow Time's Cycle*, Stephen Jay Gould offers a clear textual history of Hutton's theory.

2. It is cumbersome to repeatedly conjoin geologist and archaeologist, so in many instances when I want to include both, I use "excavator" as shorthand and also to highlight the nature of the discovery central to both geology and archaeology—digging into the Earth yields artifacts which are examined, interpreted, and finally featured in narratives that reconstruct those fragments into meaningful wholes.

3. In his discussion of an absolute ideal that exists outside time, Hegel famously closed the gap between past and present. However, Benjamin, Arendt, and most of the nineteenth-century writers who are the subject of this study differ from Hegel in seeking not a timeless absolute but rather a privileging of the present and a temporal narrative that offers preservation of life and culture.

4. Archaeology was sometimes practiced as a special area within geology; sometimes it was linked more closely to history, or the more general antiquarianism. As a discipline in its own right, archaeology was not well defined until very late in the nineteenth century. For a more extended overview of the history of archaeology, see chapter 4.

5. Many historians of science have analyzed the debates between the catastrophists and the uniformitarians: see Gillispie, *Genesis and Geology*; Greene, *Geology in the Nineteenth Century*; Hallam, *Great Geological Controversies*;

Rudwick, *The Great Devonian Controversy*; and Secord, *Controversy in Victorian Geology*. Hallam explains that catastrophists have been generally associated with fantasy, while uniformitarians have been associated with science. He goes on to show how the reality of the controversy is far more complex. For instance, he begins with the useful clarification that catastrophism is a system and uniformitarianism is both a system and a method. Uniformitarian methods could conceivably lead to conclusions that seem more properly aligned with catastrophism (30). Hallam goes on to quote Henry Thomas De la Beche, the first director of the Geological Survey, who wrote, "The difference in the two theories is in reality not very great; the question being merely one of intensity of forces, so that probably, by uniting the two, we should approximate nearer the truth" (55).

6. It is worth noting that while Arnold published this poem in 1867, he seems to have written it as early as 1850. It thus reflects concerns raised by geology rather than by Darwin who had not yet published *The Origin of Species* (1859).

7. For a brief overview of the background to nineteenth-century geology, please see chapter 2.

8. It is worth noting that scientist-writers also used the writerly strategies of revision and rereading that we more easily associate with literary figures. In *Victorian Sensation*, James Secord makes the compelling point that much of the scientific writing of the nineteenth century should be considered as serial publication: multiple editions reflecting substantive revision reveal a culture of reading, response, rereading, and revision akin to that of serial fiction.

9. It is worth pointing out that in French, the language in which both Ricoeur and Derrida wrote about the trace, the first definition of *la trace* is a foorprint or other mark left by the passage of an object or being. I paraphrase the definition from *Le Nouveau Petit Robert*: "empreinte ou suite d'empreintes, de marques que laisse le passage d'un être ou d'un objet."

10. A full discussion of Hegel, Michelet, or nineteenth-century historiography is beyond the scope of this book. For a more extended treatment of this subject see: White, *Metahistory: The Historical Imagination in Nineteenth-Century Europe*, and Momigliano, *Studies in Historiography*.

11. McClintock makes the compelling claim that those prejudices are imported back into the colonizing culture, where the working classes and women are treated as primitives. The sites I consider (Pompeii and London) here do not involve local "savages," though the working classes associated with both places are arguably treated with a disdain akin to what McClintock describes.

12. Dégerando (1772–1842) wrote *The Observation of Savage Peoples*, published in French in 1800.

13. In his study of the exotic, *Orientalism and Visual Culture*, Frederick Bohrer applies Hegel's synchronic and diachronic approaches to history to understand art history. A synchronic study "investigat[es] a cultural moment at a single time" while a "diachronic study is one that follows a track between two cultural moments" (12). Bohrer turns his attention to archaeology and argues that, "like the exoticist subject, the archaeological artifact also can be located in a horizon of expectation" (63). In other words, the observer's own position in time determines that he employs a diachronic approach: he must consider the past alongside his own present. One wonders, then, if a synchronic view of an object from another time is even possible. Even McClintock's anachronistic space exists apart from the imperial observer, and thus is paradoxically diachronic. Where material remains are concerned, synchrony is impossible, and diachrony describes the interpretation of the trace.

14. Robert D. Aguirre, Shawn Malley, and Annie E. Coombes write about empire and the archaeology of specific regions—Mexico, Assyria, and Africa, respectively. The imperial culture also made archeology at home in Britain particularly important: as A. Bowdoin Van Riper claims in *Men among the Mammoths*, "Patriotic pride and nationalism accounted for some of the growing interest in British antiquities" (23). Excavating the origins, whether those were considered to be prehistoric, Celtic, or Roman, of the British people and their empire was at once an archaeology of empire and a grim opportunity to imagine a future when they themselves might be subjects of excavation.

15. In *The Last of the Race*, Fiona Stafford offers a comprehensive study of the cultural myth of the last man, beginning in the Restoration and ending in 1900. Stafford includes a chapter on Edward Bulwer-Lytton, whose *The Last Days of Pompeii* I discuss in chapter 4, and she connects Bulwer-Lytton's desire to reach toward the past to the influence of Lyell's *Principles*.

16. "You've Kant to make your brains go round, / Hegel you have to clear them, / You've Mr. Browning to confound, / and Mr. Punch to cheer them!" (33–36).

17. In *Tess of the D'Urbervilles* Hardy famously sacrifices his doomed heroine at Stonehenge, the most widely known archaeological site in Britian, believed at the time to be a prehistoric, Celtic altar. It is a powerful location, displaying adamantly the grandeur of a long dead religion as it simultaneously suggests the coming death of the Christian religion that offers Tess no reprieve. In this anachronistic space, Hardy renders Tess an embodiment of a greatness that tragically lies ruined.

18. In *Thomas Hardy: Distance and Desire*, J. Hillis Miller offers a reading of *A Pair of Blue Eyes* that emphasizes the relationship between desire and the look. Though Miller does not address the trilobite, it seems apropos to extend his reading to Henry's meeting with the fossil: as the two gaze into each

other's eyes, they are at once brought together, Henry desiring preservation like that of the trilobite, and distanced, as stone and man will always be alien to one another.

19. It is striking that this defense should come from Buckland, who was best known in the early part of his career as a proponent of the diluvial theory of geology, which placed considerable emphasis on the biblical flood, and thus was associated more with catastrophists than with uniformitarians. In *Great Geological Controversies*, A. Hallam asserts that Buckland, along with Adam Sedgwick, was famously converted to uniformitarianism by Charles Lyell's *Principles of Geology*.

20. I am grateful to Andrew Stauffer for first suggesting the connection between this poem and my work at the MLA convention in 1999.

21. For extended discussions of Layard's archaeological work in Assyria and the removal of the bull to the British Museum, as well as Layard's popular text *Nineveh and Its Remains*, see the recent work of Shawn Malley. See also Bohrer, *Orientalism and Visual Culture*. For a description of how Layard's Assyrian artifacts were displayed in the British Museum, see Ian Jenkins, *Archaeologists and Aesthetes*.

22. Rossetti includes a footnote to explain this passage; he cites Layard's *A Popular Account of Discoveries at Nineveh* (1851), in which Layard describes the workmen's religious services, held at the excavation site.

23. Stauffer offers an important discussion of the changes Rossetti made to "The Burden of Nineveh" between 1856, when it was first published in an undergraduate magazine, and 1870, when it was published again in *Poems*. Stauffer emphasizes the revision required by a mistake made in the 1856 version of the poem: Rossetti had accidentally confused the chronology of Egypt and Assyria, a mistake which he corrected in the 1870 version. However, the error reveals, as Stauffer suggests, that the confusion of relics jumbled together has real historical consequences that go beyond accuracies of display. "Imperial hubris" leads to "confusions of history" (379).

2. The Victorian Geologist:
Reading Remains and Writing Time

An earlier version of this chapter appeared in *VIJ: Victorians Institute Journal*, volume 30 (2002) under the similar title, "The Victorian Geologist: Reading the Relic and Writing Time."

1. I paraphrase George Poulett Scrope (qtd. in Gould 1). Scrope was allied with Charles Lyell in his rejection of the diluvial theories and his resistance to forging connections between the geological record and scripture.

Scrope wrote *Memoir on the Geology of Central France* (1827), which Lyell reviewed in the *Quarterly Review*. In turn, Scrope reviewed *Principles of Geology*, and much of what we know about Lyell's intentions regarding that text comes from his correspondence with Scrope.

2. Wherever possible, I have tried to avoid gender bias in my language; however, as the male geologist-writers I consider often conflate general references to geologists with self-reference, I have determined it is most appropriate to use the masculine pronoun in reference to geologists. This choice is not meant to suggest there were not female geologists in the nineteenth century, as there most certainly were.

3. Albritton also points out that the absolute faith in uniformity was shaken in the twentieth century by the discovery that some changes were the result of catastrophic events not occurring in the present day, such as the frequent collision of meteorites with the Earth's surface.

4. Many proponents of natural theology and the argument of design remain even today, as is evident from the op-ed pages of many national newspapers in the twenty-first century. This ongoing debate reminds us of how catastrophic was the shift introduced by geology in the late eighteenth century. Writing about paradigm shifts, Kuhn describes a lengthy process wherein debate would continue well after the mainstream had accepted the new idea; two hundred years later, we seem to be still experiencing this particular paradigm shift.

5. Bishop James Ussher, writing in the early seventeenth century, had dated the creation on the eve of October 23, 4004 BCE. In *The Meaning of Fossils*, Martin J. S. Rudwick points out that at the time, Ussher's work was regarded "as a product of the finest historical scholarship" (70), and those who challenged Ussher's conclusion challenged only the particulars, not the overarching assumption that such a precise date could be determined.

6. William Buckland taught geology at Oxford, where Charles Lyell was one of his pupils. Robert Jameson taught natural history at the University of Edinburgh, where Charles Darwin attended his lectures. Buckland eventually renounced his catastrophist view after reading Lyell's *Principles of Geology* and being persuaded by the uniformitarianism so eloquently described therein.

7. Written in a letter dated May 1851 to Henry Acland (Ruskin, *Works*, vol. 36, 115).

8. In *Geology and Religious Sentiment*, J. M. I. Klaver presents a nuanced view of the struggle between science and religion. He argues that Lyell always maintained his Anglican faith and that, in fact, much of the science in *Principles of Geology* is crafted under the influence of that faith. While Klaver persuasively complicates the relationship between Lyell's religion and his science, my emphasis is on the attention to the figure of the geologist, a

figure who derives his authority from his scientific work far more than from his natural theology.

9. Note that Scrope demonstrates the parallel loyalties to theological and scientific integrity so characteristic of the uniformitarian school of geology: he praises Lyell for neither degrading scripture nor impeding scientific progress. The two stand apart from each other yet, Scrope suggests, it is imperative they coexist.

10. Richardson was not a scientist but worked closely with Mantell as the curator of his Mantellian museum. Chambers, a bookseller widely known for his popular periodical *Chambers's Edinburgh Journal*, wrote the controversial *Vestiges of the Natural History of Creation* (1844), a book that offers a rational account of Earth history and angered many who maintained a providential view.

11. I am indebted to James Secord for calling this wonderful illustration to my attention and for going to considerable trouble to send me a copy of it.

12. Greene goes on to complicate the central position in the history of geology usually assigned to Lyell. Nonetheless, here he articulates the received genealogy: Lyell crafts a vision of Earth history that gives Darwin the temporal scope required for his theory of evolution. As has been often noted (by Gillian Beer in *Darwin's Plots*, for one example) Lyell's *Principles of Geology* was one of the very few books Darwin took with him on the *Beagle*.

13. It should be noted that Lyell had favorably reviewed Scrope's *Geology of Central France* a few years earlier, and the two men were friends, so Scrope's review may be somewhat hyperbolic. Certainly, his description of the "luminous arrangement" (417) of *Principles* seems a bit extreme.

14. Secord observes in *Victorian Sensation* that John Murray published later editions of Lyell's *Principles* in a format meant to be accessible to working class readers with limited financial resources, a bit of publishing history that further confirms the success of *Principles*.

15. For histories of geology that challenge the importance of Lyell, see Rupke, *The Great Chain of History*; Greene, *Geology in the Nineteenth Century*; and Rudwick, *The Meaning of Fossils*. Other men offered as founders of geology include James Hutton, William Smith, Georges Cuvier, William Buckland, and Richard Owen.

16. Martin Rudwick's recent *Bursting the Limits of Time* (2005) offers a comprehensive history of the discovery of geological time in the late eighteenth and early nineteenth centuries. Rudwick emphasizes the revolutionary concept of prehuman time, significant for the difficulty of knowing what happened with no human witness, and shattering to a human identity formerly rooted in a Biblical narrative that offered only five days of prehuman history.

17. De Maillet wrote the *Telliamed*, published after his death in 1748, which presented a fantastical discourse from an Indian philosopher on the history of the Earth. It was widely read and is considered to have influenced Lyell and Darwin, among many others. Buffon published the more traditional *Natural History* (1749) and *Epochs of Nature* (1778): he suggested the biblical flood had no scientific importance, and he contributed to the understanding of strata. Albritton credits Buffon with "arous[ing] popular interest in natural science" (88). In addition, Giambattista Vico (1668–1744) famously offered a dynamic vision of history that required an expanded time scale, and many see Vico as the first to seriously rethink time. For more on the changing view of time see Stephen Toulmin and June Goodfield, *The Discovery of Time*.

18. Smith was a surveyor who in his work discovered the principle of faunal succession; that is, fossil remains of particular fauna always appear in the same succession within sedimentary rock. Smith most famously was the first to use fossils as the basis of geological mapping, the subject of the recent popular account *The Map that Changed the World* by Simon Winchester.

19. This is a very simplistic definition of stratigraphy, and I remind my readers that they must look elsewhere for an introduction to geology or for a comprehensive overview of its history.

20. Highly respected by many of the English scientists of the early nineteenth century, Lyell among them, Cuvier can be considered the founder of paleontology. He was also associated with the catastrophic view of Earth history because of his emphasis on natural revolutions that radically altered the Earth and its inhabitants.

21. For an impressive close reading of Lyell's rhetorical strategies, see Rudwick, "The Strategy of Lyell's *Principles of Geology*."

22. In "The Strategy of Lyell's *Principles of Geology*," Rudwick explains that many of Lyell's "scientific colleagues . . . readily accepted a vast time scale on the intellectual level, [but] Lyell seems to have recognized that it was their scientific *imagination* that needed transforming. Much of the detailed argument of *Principles* is therefore designed to draw out the full implications of a belief which they already professed" (Rudwick's emphasis; 11).

23. For extended analyses of Lyell's rhetoric in his history of geology see Klaver, *Geology and Religious Sentiment*; Rudwick, "The Strategy of Lyell's *Principles of Geology*"; and Porter, "Charles Lyell and the Principles of the History of Geology." These three separately argue that Lyell is highly selective in his account of the history of geology, and that he emphasizes the catastrophist-uniformitarian controversy and de-emphasizes uniformitarian precursors with whom he disagreed on key points, some of them theological.

24. We can see how deeply Lyell penetrated the Victorian imagination in the Megalosaurus that lumbers through the London fog in the opening paragraph of Dickens' *Bleak House*. Dickens was probably familiar with Lyell's prediction and with the humorous sketch of Lyell as an ichthyosaurus lecturing to other reptiles, which was widely disseminated. De la Beche later sold copies of the drawing to raise money for the Lyme Regis fossilist Mary Anning.

25. Jerome H. Buckley offers progress and decay as models of Victorian perceptions of change in *The Triumph of Time*. In *Time's Arrow Time's Cycle*, as the title indicates, Stephen Jay Gould considers in depth the directional models employed by geologists throughout the history of the science. He associates Lyell with time's cycle.

26. William Paley (1743–1805) published *Natural Theology* in 1802 and is associated with the eighteenth-century argument for design, the famous Watchmaker analogy. He argued that whatever is found in nature must have been designed by an intelligent creator: the more intricate the material of nature, the more intelligent the creator. Darwin directly took Paley on with his discussion of the evolution of the eye.

27. Rudwick also eschews locating Lyell in either a cyclical or a directionalist camp. He takes from Toulmin and Goodfield the term "steady-state" to describe Lyell's model of time. He explains Lyell's view: ". . . the physical features of the earth are in a state of perpetual flux. But this principle of change consists of endless fluctuation around a stable mean . . ." (1970, 17).

28. Lyell references Barthold Georg Niebuhr whose *History of Rome* (1811–1832) introduced a new and exhilarating form of historical scholarship that emphasizes close textual analysis and a stripping away of layers of artifice to uncover so-called historical truth. It is no coincidence that Niebuhr's metaphorical excavation was attractive to Lyell, as well as to many of the archaeologists I consider in chapter 4.

29. It is worth quoting Klaver at length here because he so insightfully notices this rhetorical pattern which actually stretches over an extended portion of Lyell's text (*Principles* 117).

30. I should note that we now believe the erroneously named temple was actually a market, but Lyell and his contemporaries operated under the misconception that it was a religious monument.

31. That Lyell successfully put the geologist in the position of the narrator is evidenced by the later versions of this frontispiece. In later editions, Lyell replaces the frontispiece included here with a more current drawing of the Temple of Serapis, dated 1836. In this later version, there is no figure included in the landscape. I do not consider that this compromises my reading. Rather, I argue that when he published the first volume, Lyell could not know how successful his *Principles* would be, and he relied on this visual assistance to extend his claims. Because his claims were in fact so well received, he did not need to hint at

the power of the geologist in the later editions of the book. That power was clearly evident.

32. Mantell's introductions to geology are: *The Fossils of the South Downs* (1822), *The Wonders of Geology* (1838), *The Medals of Creation* (1844), *A Pictorial Atlas of Fossil Remains* (1850), and *Petrifactions and Their Teachings* (1851). In addition, Mantell wrote *Thoughts on a Pebble*, a slim volume published in 1849 for a juvenile audience.

33. Gillian Beer notes that Darwin begins *The Origin of Species* with the words "When we look." She emphasizes Darwin's immediate evocation of observation, but she might also have pointed out that this extraordinary work begins with the scientist-observer (not species or even origins) in subject position.

34. Mantell's books, like Lyell's, alarmed natural theologians just as they empowered those willing to set aside the traditional interpretation of Earth history. This is dramatized in Thomas Hardy's *Tess of the D'Urbervilles* when the event that definitively separates Angel Clare from his father and the church is the arrival at their house of a copy of a provocative book, identified by Tim Dolin in a note as Mantell's *Wonders of Geology* (423).

35. The circumstances of this initial meeting are described in Wilson 92ff and Dean, *Gideon Mantell and the Discovery of Dinosaurs*, 41ff. Dean echoes Wilson: "Before parting, the two men had become lifelong friends" (43).

36. Dean's *Gideon Mantell and the Discovery of Dinosaurs* offers a comprehensive treatment of Mantell's career. Mantell's rise from provincial surgeon to respected author and member of the Geological Society represents a new trend in nineteenth-century society: the ability to achieve social mobility through professional, in this case scientific, successes. Unlike Lyell, Mantell was not born to be a man of letters, but his achievements as a geologist gained him access to that class.

37. We can be sure that this drawing gratified Mantell because he reproduces it with slight variation at the end of *Medals* and reprints it seven years later in *A Pictorial Atlas of Fossil Remains*.

38. It is worth connecting Mantell's two house museums to the trend in the nineteenth century to bring museum elements into every home. Barbara J. Black discusses this subject at length in *On Exhibit: Victorians and Their Museums*. She includes a chapter on the most extreme example of the house museum, the Soane Museum, which illustrates the "utopian impulse to turn every house into a museum rather than a castle . . ." (71).

39. King William IV had planned a visit in October 1832 but was unable to see the museum. His interest in Mantell's work, however, was made clear by his request to have *The Geology of the South-East of England* (1833) dedicated to him, which it was.

40. In her discussion of museum architecture and the display of natural science in the nineteenth century, Carla Yanni extends Knell's claim: "Traditional histories of science have given experiment a privileged place, but it is important to remember that in the nineteenth century, observational science was the cutting edge" (11). She regards museums as the locus and manifestation of this observational science.

41. In *A Social History of Archaeology*, Kenneth Hudson makes the compelling point that the expanding rail network also made archaeology far more accessible. He argues that trains were a valuable resource to local archaeological societies (44). A. Bowdoin Van Riper also attributes the expansion of archaeology in the 1840s, at least in part, to the accessibility of the railway in *Men among the Mammoths* (22). Accidental archaeology is the subject of chapter 4.

42. When the Great Exhibition of 1851 was closed, the Crystal Palace was moved from Hyde Park to Sydenham, between 1851 and 1854. The building remained there until 1936 when it was destroyed by a fire. Today, the park remains, including some of the original exhibits, like the dinosaur park.

43. It is worth noting that Mantell's personal writing frequently reflects a sense of paranoia. Far more important than being on the outs with the quarrymen, Mantell eventually fell out with Lyell, was ostracized by the Geological Society, and was abandoned by his family.

44. It is striking that Mantell concludes his work not with a list of fossil specimens or with a glossary of scientific terms, both common appendices to popular introductions to science like Mantell's, but rather with a catalog of dealers. The emphasis is thus on the market and on acquisition rather than on the principles or even the particulars of the science itself.

45. Waterhouse Hawkins, who constructed the life-sized dinosaur models at the Crystal Palace Park, among them the iguanodon, is the best-known craftsman of artificial relics.

46. In addition to being visited by Lyell and Mantell among many others, Anning corresponded with William Buckland, Roderick Murchison's wife, and Adam Sedgwick. She was also friendly and in frequent communication with Henry Thomas De la Beche, who was also from Lyme. There is some speculation that Anning and De la Beche may have been romantically involved, but there seems to be little to no concrete evidence to support this theory.

47. Even in accounts of Anning's life written today her worth is almost always gauged by the prices she commanded for her finds—she earned 120 guineas for a plesiosaur, for instance (Forde 14). It is interesting to note that the majority of texts published in the twentieth century that offer biographies

of Anning are written for children. She is one of a small number of women scientists to offer role models for the readers of young adult and children's biographies, and so she has become a popular subject for these texts. Even in these children's books, books that theoretically acclaim her work as a scientist, Anning's accomplishments are measured in pounds sterling.

48. In *The Hummingbird Cabinet*, Judith Pascoe explores Anning's use of literature, specifically a late poem by Byron, to contribute to her self-definition as scientist and proto-feminist.

49. Brantlinger enumerates key elements of the mid-century imperial ideology: one is the chauvinism articulated with reference to Trollope; another is the emphasis on the military.

50. Simon Knell discusses the Geological Survey and observes that it ultimately made obsolete the lone, local geologist with expertise in his own small region and led instead to the practice of geology in coordinated teams. Thus, the survey was damaging to Mantell's method of practicing geology; however, by the mid-1840s when the survey began to have an impact on the national practice of geology, Mantell had largely given up his own geological career and devoted himself instead to his writing.

51. For an extended discussion of the relationship between geology and archaeology see Torrens, "Geology and the Natural Sciences: Some Contributions to Archaeology in Britain 1780–1850." Torrens even points out that Mantell's very first publication was on archaeology. He writes, "1811 saw the first appearance in print of a significant scholar who bridged the two disciplines: Gideon Algernon Mantell (1790–1852). This 1811 publication on archaeology was his first, before he had published anything on geology, the field in which he was later to become famous. The paper announced the 'Fine Roman Pavement discovered at Bignor' in Sussex. Mantell became one of the first to make contributions to both the new geology and the new archaeology" (49). Torrens is also the greatest living authority on Mary Anning.

3. Tennyson's Fairy Tale of Science

Quotations from Tennyson's poetry are from *The Poems of Tennyson in Three Volumes* (1987), Christopher Ricks, ed, unless otherwise indicated.

1. Whenever possible I use gender-neutral language, but in referencing literature that offers "man" as something more than the male gender, I find it sometimes necessary to follow that usage for the sake of my own prose.

2. Like *The Princess*, "Locksley Hall" is about many different subjects. In its attention to science, class unrest, the marriage market, and military life, not to mention nostalgia and unrequited love, it, too, could be called a medley.

3. See especially Dean, *Tennyson and Geology* (1985) and Gliserman, "Early Victorian Science Writers and Tennyson's *In Memoriam:* A Study in Cultural Exchange" (1974–1975). Most of the scholars who address Tennyson's reading in the sciences take as a key source the catalogue of his books held in Lincoln: *Tennyson in Lincoln: A Catalogue of the Collections in the Research Centre*, vol. 1, compiled by Nancie Campbell (1971).

4. For classic discussions of *In Memoriam* and science see Dean and Gliserman. Also Millhauser, *Fire and Ice: The Influence of Science on Tennyson's Poetry* (1971), Mattes, *In Memoriam: The Way of a Soul* (1951), and Armstrong, "Tennyson in the 1850s: From Geology to Pathology—*In Memoriam* to *Maud*" (1992). Other articles and books that address this subject, particularly in relation to *The Princess*, are listed in the bibliography.

5. *In Memoriam* was written over a period of seventeen years, from 1833, when Tennyson's good friend Arthur Hallam died suddenly (Tennyson was twenty-four), to 1850, when the poem was published. *The Princess* was composed simultaneously, though not over such an extended period: Tennyson began work on *The Princess* in 1839 and published it eight years later on Christmas day of 1847. He continued to revise the poem over the next three years and the third edition, published in 1850, was heavily revised.

6. Dean lists the lyrics that were probably influenced by geology: prologue, 34, 35, 36, 39, 43, 54, 55, 56, 69, 70, 95, 112, 113, 118, 120, 123, 124, 127, 131, and epilogue (9). Eleanor Mattes also takes the lyrics one by one and charts their respective dates of composition. Mattes's data alongside knowledge of when Tennyson read particular texts in geology allows Dean to make his claims.

7. In a brief discussion of Tennyson, Nicolaas Rupke argues that Tennyson was influenced far less by Lyell than by William Buckland. See *The Great Chain of History*, 225–30.

8. For clarity, I quote here the full passage:

They say,
The solid earth whereon we tread

In tracts of fluent heat began,
  And grew to seeming-random forms,
  The seeming prey of cyclic storms,
Till at the last arose the man;

Who throve and branch'd from clime to clime,
  The herald of a higher race,
  And of himself in higher place,
If so he type this work of time

Within himself, from more to more. (7–17)

9. It is easy in reading *In Memoriam* to forget that Tennyson completed the poem almost ten years before Darwin published *On the Origin of Species* (1859). What appears to be prescient anticipation of Darwin is really a reflection of how the ideas so famously articulated by Darwin were a part of the Victorian culture, presented in more or less tentative form by the geologists, among them Charles Lyell, and more or less scandalous form, like Chambers's *Vestiges of the Natural History of Creation*. This pre-Darwinian discourse enabled Darwin to make such an extraordinary impact and prepared his readers to understand and accept his theories.

10. Tennyson also explores the conjunction of marriage and geology in "Locksley Hall," but in this 1842 poem, marriage fails to rescue the speaker from heartache or from the withering oppression of geological evidence. It is worth noting that Matthew Arnold also offers love, possibly marriage, as a shield against the cruelty of geology's nature in "Dover Beach" in which his speaker cries, "Ah, love, let us be true to one another" for the Earth has proved false to us all. See chapter 1 for a more extended discussion of "Dover Beach."

11. We see a similar sort of temporal medley in some of the writing inspired by archaeology, the subject of the next chapter. For instance, in *The Last Days of Pompeii*, Bulwer-Lytton interjects descriptions of the ruins at Pompeii alongside his fictional account of those places in their heyday.

12. Johnston reads *The Princess* as extolling the notion of self as medley (550). She describes the poem as a fairy tale, arguing that Tennyson employs the trope of splitting: Ida's characteristics, for example, are parceled out to her female companions, Psyche as mother and Blanche as patriarch. The ideal exists in the marriage of the prince and the princess when the medley will be contained within a single union. Other critics have considered such splitting from a gendered perspective: the prince is both male—he saves the princess from a raging river—and female—his "weird seizures" (Stone).

13. Killham goes on to examine Tennyson's representation of relationships and gender roles and argues that the tone is actually quite earnest.

14. Herbert argues that the frame serves to control polyphony. She claims that the poem addresses the threat of true democracy and finally attempts to stifle "social and linguistic freedom within its monophonic voice" (146). I take Herbert's point further: the monophony of the supposed medley reveals Tennyson's assertion of a single narrator over the competing voices of geological debate and, more importantly, over the alarming absence of any narrative authority in the history of the Earth.

15. Using a Mechanics' Institute festival to introduce science in *The Princess* directly links science with women's issues. Killham explains that Mechanics' Institutes tended to be forums for the Owenites, who espoused a progressive, feminist ideology, and, with their festival, Tennyson invoked this new

feminist ideology. Tennyson's readers would have associated women's education and rights with those of the working class and may have been familiar with the Owenites' efforts to educate the public about social reforms, especially pertaining to women and marriage: in short, they would easily have associated the festival in the poem with ideas about reforming women's role in society. The festival thus serves as a backdrop not only to the events within the poem but also to the thematic conjunction of women's issues and science.

16. While it is easy to elide the house museum and the scientific festival, it is important to note some key differences between them. For instance, the house museum is a permanent display while the festival is by definition temporary. The museum is accessible only to those invited into the house, and is thus more elite; the festival is open to the public and has the express purpose of making available to all classes this temporary display. In *The Birth of the Museum*, Tony Bennett describes museums and festivals as opposites, the one an embodiment of order and the other a barely controlled chaos. This stark contrast is not appropriate in *The Princess* because the house museum is not a public institution and, in this case, does not seem to be well ordered. (Indeed, in the 1840s, Tennyson's readers would have recognized how very different from a formal, chronologically ordered museum is this haphazard collection of curiosities, displayed for their singularity rather than for any collective value.) It is useful to recall that though the museum and the festival coexist in Tennyson's setting, they are not to be conflated.

17. The first institution of higher learning for women, Queen's College suffered from mismanagement, but it nonetheless paved the way for the renowned women's colleges Bedford College, founded in 1849, and Girton College, founded in 1869. Dorothea Beale, an activist and professional in the world of women's education, offers a firsthand account of the early years of Queen's College and the education reform movement in general in "Girls' Schools, Past and Present."

18. This overview of women's education in the mid-nineteenth century reduces the issue to a most elementary level. It is not my intention to summarize the terms of the complex debates on this matter. Rather, I intend to show how Tennyson could possibly have drawn a parallel between such seemingly unrelated topics as education for women and geology. For thorough treatments of the movement for women's education see Burstyn, *Victorian Education and the Ideal of Womanhood*, and Hunt, *Lessons for Life: The Schooling of Girls and Women, 1850–1950*.

19. I should remind my reader that I offer a literary analysis, not a scientific or a historical one. Psyche is a fictional character created by a fictional character created by a poet. None of these figures is a scientist. Nor am I. Nonetheless, they voice the concerns of a generation for whom science changes the world, as do I.

20. Succession is a precursor to evolution. Epochs follow after earlier epochs, and one can read this geological succession as progression, each epoch offering a more advanced species. Succession differs from evolution in that succession describes a history but has nothing to say about species changing, or evolving.

21. Scrope published *Memoir on the Geology of Central France* in 1827. Scrope and Lyell were friends and, through reviews, popularized each other's works. Stephen Jay Gould identifies the passage quoted here as a cliché in geological textbooks: I quote from his *Time's Arrow Time's Cycle* in which he uses the quote as an epigraph.

22. In his note "Tennyson's *Princess* and *Vestiges*," Milton Millhauser reads these lines as a description of Chambers's take on the nebular hypothesis: a single act of creation set all other creative acts in motion. The explicit reference to "succession" supports this reading.

23. See chapter 1 for an extended discussion of the trace.

24. And the person he does describe is a masculine one. The carefully misplaced modifier in line 409 suggests that in her youth, Ida was a boy.

25. Ida feminizes his strength, attributing it to their mother: "whereas I know / Your prowess, Arac, and what mother's blood / You draw from, fight; you failing, I abide / What end soever" (5:393–96).

26. In the second stanza, which I do not quote for the sake of space, we learn that some of the bugles blown are the "horns of Elfland" (10). The poem thus describes a fading musical conversation between humans and the dying elves. It is striking to consider the analogy between the fading away of fantastical creatures and the sense in the nineteenth century that humans as we have always known them are also dying out, or that, like elves and dinosaurs, humans will one day be nothing more than legendary or fossil traces.

27. Killham argues that Tennyson responds to Caroline Norton's infamous custody battles in *The Princess*, and we see this reading most clearly in the transfer of Psyche's infant from one parent to another.

28. Sarah Ellis is the author of the exceptionally popular *Women of England* series that offered guidance to women in assorted relationships in how to behave, especially in relation to the men in their lives. Chase and Levenson consider the wonderfully odd triad of Ellis, Caroline Norton, and Queen Victoria, all women who represented distinct versions of women's power. Princess Ida is a medley of all three women. Chase and Levenson also discuss Florence Nightingale, but she had not yet become the lady with the lamp nor had she written *Notes on Nursing* when Tennyson wrote *The Princess*. His nurses are thus not the proto-professionals of Nightingale's tradition but are rather Ellis's ideal wives and mothers who rely on instinct and affection in their care of the wounded.

4. Accidental Archaeology
in London and Pompeii

1. The explosion of Grecian artifacts inspired such proclamations as Keats's famous, "Beauty is truth, truth beauty." Though the line has become a cliché, it usefully reminds us how firmly associated in the nineteenth-century psyche were the art objects of the classical world and an idealized sense of culture, beauty, and truth, qualities rarely associated with artifacts from other sites, such as Nineveh or lesser Roman cities, including Pompeii and Londinium. When Austen Henry Layard brought the Nineveh sculptures back to London, for instance, the curators of the British Museum did not know what to do with them. They were, as Jenkins says, the archaeological equivalent to Darwin's "missing link," showing the progression from primitive to more civilized and certainly important, but as certainly inferior.

2. I will use the Latin name for London to distinguish the Roman city from the Victorian one and also to highlight archaeological excavation of Roman remains, as opposed to medieval, prehistoric, or any other sort of remains in London.

3. Hegel's original history has to do with total immersion in the historical moment; reflective history involves seeking relation between past and present; and philosophical history considers that everything is present—there is no past. Vis-à-vis Pompeii and London, I would classify most of the histories as reflective—they emphasize the likeness between past and present and so close the gap. Hayden White's *Metahistory* offers a thorough discussion of competing historiographies in the nineteenth century.

4. Vance refers to Macaulay's *Lays of Ancient Rome* (1842), just one example of a popular text that celebrates Rome, even as interest in that classical empire waned. Vance's treatment of the Victorians' relationship to ancient Rome is comprehensive. For further discussion of the Victorians' view of the classical world, see Richard Jenkyns, *The Victorians and Ancient Greece* (Harvard University Press 1990) and also Jenkyns, *Dignity and Decadence: Victorian Art and the Classical Inheritance* (Harvard University Press 1992).

5. For a comprehensive treatment of these societies and their role in the definition of archaeology, see Levine, *The Amateur and the Professional: Antiquarians, Historians, and Archaeologists in Victorian England, 1838–1886.*

6. Though most historians of archaeology dismiss the work of the first half of the nineteenth century as unscientific, A. Bowdoin Van Riper is an exception: "Presidents of the national archaeological societies repeatedly emphasized the value of empiricism in their addresses at annual meetings. The proper business of the archaeologist was the collection, comparison, and organization of facts; once enough facts had been gathered, they would somehow fuse into an accurate, impressively detailed picture of the past" (35).

These demands for empiricism, however, seem to have come without any pre-scribed methodology. That is, early archaeologists may have appreciated the need for a systematic approach to a site, but there was little agreement about what system would most accurately reveal the past.

7. Stukeley also studied Roman remains in Britain. In 1722, he and some friends formed a club for the study of Roman Britain, called the Society of Roman Knights, a name that suggests Stukeley also took a Romantic approach to Roman Britain.

8. Stuart Piggott has written extensively on William Stukeley and pro-vides a thorough biography that examines Stukeley's archaeological work and interest in druids within the context of the Romantic movement. This reading of Stukeley sheds light on one aspect of antiquarianism: the Romantic attrac-tion to ruins as a sublime manifestation of the integral relationship between nature and the built environment. See Piggott's biography *William Stukeley* and his more general histories of antiquarianism, *Ruins in a Landscape* and *An-cient Britons and the Antiquarian Imagination*. David Boyd Haycock challenges Piggott's reading of Stukeley.

9. A particularly delightful explanation for stone arrowheads is that they were made by fairies, an explanation more probable to many in the early nine-teenth century than the alternative, that humans lived hundreds of thousands of years ago.

10. Lyell offered the conservative figure of 100,000 years for the age of humankind (Daniel 118), but even this number, which we now know to fall far short of the actual age of *homo sapien* (approximately 400,000 years old and distinct from other species of hominid that stretch back millions of years), represents an enormous leap from the previous estimates of 6,000 to possibly 8,000 years.

11. Indeed, it was in 1860, just after Lyell addressed the British Associa-tion, that the excavations at Pompeii were revolutionized and made much more empirical and methodical.

12. Hugh Miller was a noted Scottish geologist and publisher, perhaps best known for *The Old Red Sandstone* (1841) and *The Testimony of the Rocks* (1857). Miller is an interesting figure, for though he devoted himself most consistently to geology, he also demonstrated a keen interest in folklore and local archaeology. For an overview of all aspects of Miller's life and career, see *Celebrating the Life and Times of Hugh Miller*, edited by Lester Borley.

13. For an extended and illuminating analysis of Shelley's sonnet in the context of Romantic antiquarianism, see Leask, *Curiosity and the Aesthetics of Travel Writing, 1770–1840*. Leask examines potential sources for "Ozyman-dias," both literary and artifactual, and offers a reading of the poem that ex-plores Shelley's engagement with complicated issues of ruin, but also empire, aesthetics, and curiosity in broad terms.

14. Archaeology can be divided into two major areas of exploration: the study of literate societies and the study of preliterate or prehistoric people. Rome, of course, was a literate society, and thus the attention to prehistory that occurs at the border between geology and archaeology may seem removed from my study here. However, the interest in Pompeii and Londinium had to do with the ways in which those sites reflected the present, rather than in what their artifacts revealed about the past (how they may have corroborated or complicated the textual evidence), so the literacy of the Roman people and whatever records they may have left behind is merely incidental to my analysis, which emphasizes encounters with remains, not history.

15. Glyn Daniel considers that archaeology developed as a scientific discipline with the rise of prehistory in the 1850s, and, indeed, a marked shift in the approach to Pompeii occurs at mid-century. To support his dating of archaeology's beginnings, Daniel notes that the Great Exhibition of 1851 had no prehistory exhibit, whereas the 1867 exhibition in Paris did (115). Daniel considers some of the excavations that took place at Pompeii at the start of the nineteenth century to have been advanced for the time. He writes: "Right at the end of the eighteenth century and at the beginning of the nineteenth, due to the keen and generous interest taken by the Napoleonic kings of Naples, carefully planned excavations of Pompeii took place—the first large planned excavations in history. The plan was due to a scholar of Naples, Michele Arditi, and enough money was forthcoming to employ at times six hundred men at this work" (153). What makes these excavations noteworthy is the sheer number of workers employed at the site. Nonetheless, excavation at Pompeii was only sporadically scientific until Giuseppe Fiorelli took over in 1860.

16. Even Richard Colt Hoare (1758–1838) and William Cunnington (1754–1810), whom Marsden offers as the first archaeologists to apply scientific methods to barrow digging, used working techniques that were, as Marsden puts it, "by present standards . . . execrable" (16).

17. The ability of the archaeologist to effortlessly destroy the remains of past civilizations may be scientifically accurate but it also reflects the imperial ideology of the time: the cultural superiority of the archaeologist goes unquestioned, and his easy destruction of bodies and artifacts is not surprising.

18. Of 20,000 inhabitants, fewer than 2,000 were killed by the volcanic eruption. Ten percent is certainly a significant portion of the population, but when we consider the complete absence of any sort of advance-warning technology, beyond the naked eye, it is remarkable that more Pompeians were not victim to the blast. Even in twenty-first century representations of the disaster at Pompeii, such as the 2004 Discovery Channel docudrama, the emphasis is on those who died and the exaggerated possibility that an even worse catastrophe is imminent.

19. Similarly, in "A Day at Pompeii," the author offers a dream in which he was at Pompeii at the time of the eruption and so narrates the event in the

first person. He, too, describes the plume of smoke: "It soon reached an elevation of, I should judge, nine thousand feet . . . it flattened out at the top like a spread umbrella or the branches of an Italian pine . . ." (741).

20. It should be noted, however, that Thackeray did not think much of Bulwer-Lytton, and this passage from *The Newcomes* is ironic.

21. See chapter 1 for discussion of these geological reconstructions. For the reconstructions themselves, see Rudwick's *Scenes from Deep Time*.

22. Middleton seems to have published nothing other than this poem, and the dates of his life are unknown.

23. John Delaware Lewis later translated the letters of Pliny the Younger. Thus, Lewis knew well both the material and textual histories of Pompeii.

24. Yet, as I note above, their attention is held almost entirely by the Pompeians of the past. The living Pompeians hold no interest in themselves, and are only noteworthy insofar as they make the dead city appear to be a living one.

25. Very few accounts of Pompeii attend to the ruined city as a symbol of the Roman Empire. Because Pompeii was essentially a resort, it is sometimes considered symbolic of the excesses of the wealthy, and its destruction is seen as a punishment for vanity. We see this in Bulwer-Lytton's *The Last Days of Pompeii* with the character of Julia. However, most descriptions of the city and fictional accounts of life therein highlight the lives of regular men and women, such as soldiers, street sellers, and children.

26. While nineteenth-century visitors traipsed carelessly over the mosaic floor, it has since been protected with a small fence, presumably to preserve the mosaic, so we can imagine that some modicum of privacy has been restored to this Pompeian home.

27. See Chase and Levenson *The Spectacle of Intimacy* for a rich treatment of this Victorian phenomenon of broadcasting the private.

28. I am grateful to Alison Booth for first suggesting this connection.

29. The graffiti also stand in marked contrast to Pliny's more objective, formal text, and were perhaps particularly appealing because of the contrast.

30. A possible model for the Pompeian Court was the exhibition of Egyptian antiquities held at the Egyptian Hall in Picadilly in 1821–1822. This exhibit featured a full-scale model of a tomb, and visitors were free to walk around inside the "tomb."

31. It seems especially appropriate that the Pompeian Court should have been intended as a site for refreshment; this use confirms that Pompeii seemed familiar and comfortable.

32. For extended discussions of metropolitan improvements in Victorian London, see Olsen, *The Growth of Victorian London*, and Nead, *Victorian Babylon*.

33. Giovanni Battista Belzoni famously excavated the tombs of ancient Egypt in the early nineteenth century; he and his finds became famous in 1821 when his Egyptian discoveries were exhibited at the Egyptian Hall in Picadilly. The year before, John Murray published Belzoni's *Narrative of the Operations*, which was very popular and further contributed to Belzoni's fame. His finds also played an important role in the establishment of the antiquities collection in the British Museum. For a brief description of Belzoni's contribution to the British Museum, see Miller, *That Noble Cabinet*. For an analysis of Belzoni's exhibition in the Egyptian Hall and its significance, see Pearce, "Giovanni Battista Belzoni's Exhibition of the Reconstructed Tomb of Pharaoh Seti I in 1821" published in the *Journal of the History of Collections*. See Leask, *Curiosity and the Aesthetics of Travel Writing, 1770–1840*, for a vivid description of Belzoni's career, personality, and rapacious proto-archaeology. For a more extended treatment of the fascination with Egypt in general, see Curl, *Egyptomania*.

34. The purchase of Roach Smith's collection was a key step in the creation of the British Museum's native archaeological collection, a priority in the 1850s. For a discussion of this acquisition, and for a glimpse into Roach Smith's "belligerent" personality (134), see Wilson, *The British Museum: A History*. For a more general discussion of the development of the British Museum's antiquities holdings, see Miller, *That Noble Cabinet*.

35. Roach Smith was not alone in recognizing the incidental character of building sites as also archaeological excavations. In his biography of one of the most famous archaeologists of the later nineteenth century, *Pitt-Rivers*, Mark Bowden recounts: Pitt-Rivers's "attention was drawn to the discovery of archaeological remains on a building site at London Wall by an article in the *Times* on 20 October 1866 describing the removal of twenty cartloads of bones from the site. He visited the site the same day and found a large excavation being made for an extension to a wool warehouse. He maintained a watching brief on the site for at least two months" (64). Glyn Daniel and Colin Renfrew also offer a compelling example of archaeology in London: in *The Idea of Prehistory*, they assert that the first known discovery of human remains alongside remains of extinct animals was in Gray's Inn Lane in London at the end of the seventeenth century (31).

36. Pitt-Rivers echoes Roach Smith's condemnation of the ignorance that leads to the destruction of valuable archaeological sites. Bowden quotes Pitt-Rivers from the *Anthropological Review* (1867): "Thus while the remotest parts of Europe are being searched for the vestiges of lake dwellings, and the most valuable reports on the same subject are received from the four quarters of the globe, similar remains are in daily process of destruction at our own doors by persons who are ignorant of their meaning and of the importance that attaches to them. This certainly ought not to be" (65).

37. Kenneth Hudson discusses the struggle between forward and backward movement that always complicates archaeology, but he fails to notice that struggle in the earlier part of the century: "The Victorians of the 1870s were the first generation to be seriously faced with the always difficult task of arbitrating between rival claims of history and tradition on the one hand and progress and a steadily improving standard of living on the other" (55). My presentation of Roach Smith's polemics confirms Hudson's point but reveals that he should have reached farther back in the history of archaeology.

38. Benjamin here echoes Hegel.

39. One of the most important late Victorian works on degeneration is Max Nordau's *Degeneration* (1895). Twenty years earlier, Cesare Lombroso published *Criminal Man* (1876), which offered the hypothesis that lower classes were literally a lower order of human.

40. Harrison's text is not the only one along these lines. Another is *Our Own Pompeii: A Romance of Tomorrow* by Samuel Middleton Fox.

41. Harrison alludes to Edward Bellamy's *Looking Backward, 2000–1887*, published in 1888. Similar to William Morris's *News from Nowhere*, *Looking Backward* is a science fiction utopia in which the future is free from war and strife.

42. Harrison's inclusion of a recording of *The Princess* highlights the fact that his Pompeian collection is also a medley intended in its diverse quality to represent the age. See chapter 3 for an extended discussion of *The Princess*.

43. Three years after "A Pompeii for the Twenty-Ninth Century," Harrison published *Annals of an Old Manor House* (1893), which represents a similar perspective on history. He excavates the layers of history at the site of his family estate, built in the sixteenth century. As he proposes regarding the English Pompeii, the old manor house conflates public and private history.

5. Dickens among the Ruins

1. For an extended discussion of writing about scientific developments in *Household Words*, see Metz, "Science in *Household Words*: 'The Poetic . . . Passed Into Our Common Life'" *Victorian Periodicals Newsletter* 11, no. 4 (1978): 121–33.

2. The Antiquaries Club had broad interests, among them not only textual scholarship but also study of material remains discovered in England.

3. See Ackroyd, *Dickens*. In July 1865, the month in which appeared the fifteenth number of *Our Mutual Friend*, Dickens wrote to Owen, and he references him explicitly in the second chapter of the novel. Dickens traveled in Italy with Layard and worked with him on administrative reform matters.

4. Similarly, Hayden White describes "the imposition of a certain formal coherence on a virtual chaos of 'events'" (qtd. in Beer 1986, 75).

5. Bleeding Heart Yard and Holywell Street are both in the neighborhood of the Strand.

6. Recall the atavism Anne McClintock describes in *Imperial Leather*. Bleeding Heart Yard is a perfect anachronistic space, where the inhabitants live not only in past time but even in a less civilized culture.

7. I discuss the comparison between Arthur and the archaeologist Belzoni later in this chapter. In her discussion of Egyptian exhibition and *Edwin Drood*, Hyungji Park describes Belzoni as an engineer—he used "primitive engineering methods as well as brute strength, bribery, and ingenuity, to plunder various Egyptian antiquities for Britain" (536)—and we thus see that Arthur's work in an engineering firm is possibly connected to his "archaeological" work. Such a connection adds meaning to the presence of the factory at the edge of Bleeding Heart Yard.

8. In an unpublished essay, Andrew M. Stauffer offers a wonderful discussion of Dickens's reflections on Egypt in London. Stauffer begins with Dickens's "Down With the Tide," published in *Household Words* in 1853, and its delightful description of the "vapour" of London containing bits of "mummy-dust" and "camels' foot-prints."

9. See chapter 4 for a general introduction to Belzoni. For an extended treatment of Dickens's interest in and use of Belzoni, see Hyungji Park, "'Going to Wake Up Egypt': Exhibiting Empire in *Edwin Drood*," published in *Victorian Literature and Culture*. Park suggests that Dickens implies a parallel between the potential marriage of Edwin Drood and Rosa and the real marriage of Belzoni and his wife Sarah. She extrapolates from a brief reference and does so convincingly. In my analysis of *Little Dorrit*, I do the same: the reference to Belzoni is brief, but following Park, I believe it to be significant, and I will make much of it. Park also notes that the head of Ramses II, one of Belzoni's major finds, was the inspiration for Percy Shelley's poem "Ozymandias," and it thus seems fitting that Dickens should invoke Belzoni in *Little Dorrit*, a novel that is so much about decay, ruin, and vain attempts at immortality.

10. Dickens also facilitated the popularization of Pompeii and the fantastic musings on London's Pompeii-like future: he published "London in AD 2346" (1846) in *Punch* and John Delaware Lewis's "The City of Sudden Death" and "Preservation in Destruction" (both 1852 and all discussed in chapter 4) in *Household Words*.

11. Nigel Leask argues that Layard had Belzoni in mind when he prepared his own travel account (155).

12. Michael Cotsell notes that Dickens and Layard climbed Vesuvius together during this trip (1986, 193). Thus, we can even connect Layard with Dickens's view of Pompeii.

13. Trevelyan (1807–1886) was a career civil servant; Northcote (1818–1887), also a civil servant, was a conservative MP.

14. The preface to the 1857 edition of *Little Dorrit* is reprinted in the 1985 Penguin Classics edition cited throughout this chapter.

15. Jonathan Smith discusses Dickens's probable allusion to Darwin's work on barnacles at length in "Darwin's Barnacles, Dickens's *Little Dorrit*, and the Social Uses of Victorian Seaside Studies."

16. Andrew Stauffer observes that paper at mid-century was often associated with its origin in rags (even in mummies) and thus had an inherent archaeological association. Dickens explores this most explicitly in *Bleak House*.

17. Mr. Clennam's lover was a pupil of Frederick Dorrit's, and Mr. Clennam specified in his will that a sum of money be left to Frederick's youngest daughter or, if he had none, to the youngest daughter of his brother. Of course, this codicil, suppressed by Mrs. Clennam, makes Amy Dorrit Mr. Clennam's heir. It is striking, however, that though seeking a concealed connection between his family and the Dorrits had been Arthur's aim, this obscure connection is overwhelmed by the more sordid matter of Arthur's true parentage and his adoptive mother's cruelty. It is significant that none of this information ever becomes known to Arthur.

18. Writing about the influence of the first and particularly the second law of thermodynamics on Dickens, Jonathan Smith notes that the term entropy was not in use in the nineteenth century, though the concept it describes worried many Victorians, Dickens among them. Smith describes the second law thus: "The second law, first formulated in Britain by [William] Thomson in 1852, asserts that in any expenditure of mechanical energy a certain amount of that energy is forever 'lost' as heat. Each transformation of mechanical energy causes heat to dissipate into the surroundings. . . . Over time, randomness and disorder increase until the universe reaches a uniform temperature, a point where motion would cease and life would come to an end" (1995, 38–39). The dissipate behavior of Fanny, Tip, and even William Dorrit can be connected to this scientific understanding of energy and dissipation.

19. The description of Meagles's collection in some ways echoes the house museum central to the prologue in *The Princess*. See chapter 3.

20. It's worth noting that the Birmingham manufacturer of faux artifacts himself asserts an interesting agency over the past. See the discussion of Waterhouse Hawkins et al. in chapter 2.

21. I argue in chapter 4 that Victorians more readily connected ruin and their contemporary society than is generally assumed.

22. The conclusion of *Little Dorrit* also anticipates Eliot's *Middlemarch*, which novel also finds a post-geological happy ending in uniformitarianism: Dorothea's "full nature, like that river of which Cyrus broke the strength, spent itself in channels which had no great name on the earth. But the effect of her being on those around her was incalculably diffusive: for the growing good of the world is partly dependent on unhistoric acts; and that things are not so ill with you and me as they might have been, is half owing to the number who lived faithfully a hidden life, and rest in unvisited tombs" (Eliot 785).

23. Throughout my discussion of *Our Mutual Friend*, I will refer to this character as John Rokesmith. I prefer his assumed name for two reasons: first, because it is the name by which we know him for most of the novel, and second, because "roke" suggests smoke or vapor and thus connotes the difficulty of knowing, a central theme in the novel, as I will discuss.

24. There are too many discussions of the significance of the dust to list them all here. Select works that have impressed me are those by Michelle Allen, Joel Bratten, and Howard Fulweiler. Please see the bibliography for an extended list of sources on *Our Mutual Friend*.

25. As discussed in chapter 4, Roach Smith was best known as an archaeologist and authority on Roman London. Charles Dickens is listed among the subscribers to his *Illustrations of Roman London*, published in 1850. Roach Smith wrote of Dickens: "That there was much good in him, thousands can testify, and thousands yet to come will be evidence to his benevolence" (98).

26. Dickens's dealings with the dust yard at Nova Scotia Gardens seem to have been less extensive and less positive than were his dealings with Henry Dodd. I mention this second instance here to demonstrate that Dickens's experience of these sites was not restricted to only one example, and that he had at least two distinct encounters with dust yards that would each have led him to imagine an inherent connection between such places and progress. I discuss Dickens's connection to Nova Scotia Gardens at more length below.

27. Of course, the lowest classes had no choice but to live in the neighborhood of the dust mounds, and so along with so many other luxuries unavailable to them, the poor also lived their lives free from the illusion that life produced no waste. "The dust-yards . . . sometimes occupy . . . open spaces adjoining back streets and lanes, and surrounded by the low mean houses of the poor" (Mayhew 229). The waste even became symbolic of the squalid living conditions, vice, and corruption that characterized the lives of so many of London's poor. The waste of the dust mounds represented not only the conditions of their lives but represented even the poor themselves.

28. There has been considerable debate over whether or not dustheaps at mid-century contained human excrement. The question matters to many critics because the presence of human waste amplifies the symbolism of the dustheaps in regard to the corruption of wealth. However, as my interpretation emphasizes the work involved with the dustheaps rather than their substance, I consider this question immaterial to my analysis.

29. Burdett-Coutts is most often connected with Dickens through their joint interest in the Urania House for homeless and fallen women: Burdett-Coutts was the house's patron and Dickens its manager. Dickens never had any formal connection to Columbia Square, but he was informally involved through his friendship with Burdett-Coutts. He surveyed plans and visited the site on her behalf in 1852. Dickens wrote to Burdett-Coutts, "The enclosed Nova Scotian proposals seem to me to be fair enough. I had previously gone over the plan carefully, with Mr Hardwick, on the ground" (March 16, 1852). It is worth noting that the construction of the project was long delayed by a condition of the sale of the land that stipulated the owner of the dust mounds had seven years to sell off the mounds. During this period, Dickens separated from his wife, which precipitated the end of his relationship with Burdett-Coutts, who sided with Catherine Dickens. He thus became distanced from the Columbia Square project, though it is clear he remained well informed about the progress at the site and certainly maintained an interest in the reform project ongoing there. In the long removal of the dust mounds and in the miserly qualities of their owner, we can see a possible source for Harmon and his yard.

30. As several critics have observed, *Our Mutual Friend*'s redemption plot is flawed. Boffin is an exceptionally good character to begin with, and his apparent corruption by the wealth he gains from the dust proves to be a ruse. His character does not evolve: most importantly, he is changed neither for the better nor the worse by his acquisition of the dust mounds. The characters who could most easily be said to have been redeemed by the novel's end are Bella Wilfer and Eugene Wrayburn, both of whom are motivated by love to give up money. Thus, the redeeming agent, if there is one, appears to be love, which has little if anything to do with the dustheaps.

31. Sicher observes that the most explicit connection between an individual and a geological relic is described in Mrs. Podsnap: "fine woman for Professor Owen, quantity of bone, neck and nostrils like a rocking-horse, hard features, majestic head-dress in which Podsnap has hung golden offerings" (21). Sicher emphasizes the reference to Sir Richard Owen, the celebrated paleontologist. Dickens thus invites his reader to consider an implicit connection between Mrs. Podsnap and the Megalosaurus of the opening paragraph of *Bleak House*. Sicher does not note the latter part of the description in which Dickens shifts from paleontology to archaeology: Mrs. Podsnap is presented

as a relic both geological and archaeological, in both cases ancient and hardened. The conflation of these three sciences—paleontology, geology, and archaeology—which we now regard as wholly distinct, would have been commonplace and unremarkable to Dickens.

32. I am grateful to Michelle Allen for her reading of Lizzie's redemption: Lizzie travels away from the tainted urban river, and in the pastoral setting she is able to apply her skills as a scavenger to save Eugene, seeking not the financial gain that characterized her father's work but instead the altruistic aim of saving a man she loves who she believes can never marry her. She is rewarded by marrying the man she loves after all, and thus ascending the social ladder considerably.

33. Metz goes on to emphasize Dickens's attempt to invest the small with value: "The idea that everything is of potential value, that nothing is so trivial, vulgar, or superficially unlovely that the imagination cannot reanimate it and make it new, is at the heart of *Our Mutual Friend*" (1979, 68). The attempt to rescue the small, the individual from insignificance, from being overwhelmed by the expanse of time and the utter indifference of nature characterizes much literature of the nineteenth century.

34. Metz confirms, "'Articulation' is the dominant metaphor of *Our Mutual Friend*, introduced with the Venus subplot, but clearly related to the struggles of various characters to sort out and unify the separate strands of their experience" (1979, 62).

35. Fulweiler offers the intriguing possibility that the French gentleman may allude to M. Boucher de Perthes, whose insistence on the presence of human remains in gravel pits was the subject of an 1861 article in *All the Year Round*. Such an allusion complements nicely the final resting place of Headstone and Riderhood.

36. Paul Hamilton paraphrases Stephen Bann who argues in *The Clothing of Clio* that "we must beware of thinking of the historian as a taxidermist. The stuffed animal of the past may appear, incontrovertibly, to be the thing itself; but such history's lack of a living relationship with our present is more of a disadvantage than the new shape, altered to fit modern needs and prejudices, in which it can continue to live with us" (23). Venus is literally a taxidermist, but later in the novel he abandons the muddle of his life and adopts a more orderly and, importantly, a more traditional domestic life. In Hamilton's terms, Venus attends finally to living relationships, and for this he is rewarded.

37. The first law of thermodynamics suggests that the total energy remains constant when a system undergoes a change, whereas the second law is associated with entropy and dictates how energy can be transferred between different components of a system.

38. I am indebted to Cindy Wentzler, a graduate student at Bucknell University, for her thoughts on this subject.

39. Recall Michelet's association of history and resurrection.

40. Mortimer Lightwood is the novel's other excavator. Many characters put together pieces, but only a few unearth fragments as Rokesmith and Lightwood do.

Bibliography

Abrams, M. H., ed. *The Norton Anthology of English Literature*, vol. 2. New York: Norton, 1993.

Ackroyd, Peter. *Dickens*. New York: HarperCollins, 1990.

Adams, James Eli. "Woman Red in Tooth and Claw: Nature and the Feminine in Tennyson and Darwin." *Victorian Studies* 33 (1989): 7–27.

Adams, W. H. Davenport. *The Buried Cities of Campania; or Pompeii and Herculaneum, Their History, Their Destruction, and Their Remains*. London: T. Nelson and Sons, 1868.

Aguirre, Robert D. "Agencies of the Letter: The Foreign Office and the Ruins of Central America." *Victorian Studies* 47 (2004): 285–96.

———. "Exhibiting Degeneracy: The Aztec Children and the Ruins of Race." *Victorian Review* 29 (2003): 40–63.

Albritton, Claude C., Jr. *The Abyss of Time: Changing Conceptions of the Earth's Antiquity after the Sixteenth Century*. San Francisco, CA: Freeman, Cooper, 1980.

Allen, Michelle. "Cleansing the City: Geographies of Filth and Purity in Victorian London." PhD diss., University of Virginia, 2001.

Arendt, Hannah. *Between Past and Future*. New York: Viking, 1961.

Armstrong, Isobel. "Tennyson in the 1850s: From Geology to Pathology—*In Memoriam* (1850) to *Maud* (1855)." In *Tennyson: Seven Essays*, edited by Philip Collins, 102–40. New York: St. Martin's, 1992.

Arnold, Matthew. "Dover Beach." 1867. In *The Major Victorian Poets: Tennyson, Browning, Arnold*, edited by William E. Buckler, 617–18. Boston: Houghton Mifflin, Riverside Editions, 1973a.

———. "Stanzas from the Grande Chartreuse." 1855. In *The Major Victorian Poets: Tennyson, Browning, Arnold*, edited by William E. Buckler, 617–18. Boston: Houghton Mifflin, Riverside Editions, 1973b.

Auerbach, Erich. *Mimesis: The Representation of Reality in Western Literature*. Translated by William R. Trask. Princeton: Princeton University Press, 1953.

Axton, William F. "Religious and Scientific Imagery in *Bleak House*." *Nineteenth-Century Fiction* 22 (1968): 349–59.

Bakewell, Robert. "A Visit to the Mantellian Museum at Lewes." *Magazine of Natural History* 3 (1829): 9–17.

Bal, Mieke. "Telling Objects: A Narrative Perspective on Collecting." In *The Cultures of Collecting*, edited by John Elsner and Roger Cardinal, 97–115. London: Reaktion Books, 1994.

Barthes, Roland. *Michelet*. Translated by Richard Howard. New York: Farrar, Straus and Giroux, 1987.

Baudrillard, Jean. "The System of Collecting." Translated by Roger Cardinal. In *The Cultures of Collecting*, edited by John Elsner and Roger Cardinal, 7–24. London: Reaktion Books, 1994.

Beale, Dorothea. "Girls' Schools, Past and Present." *The Nineteenth Century* 23 (1888): 541–54.

Beer, Gillian. *Darwin's Plots: Evolutionary Narrative in Darwin, George Eliot and Nineteenth-Century Fiction*. London: Routledge and Kegan Paul, 1983.

———. *Open Fields: Science in Cultural Encounter*. Oxford: Oxford University Press, 1996.

———. "Origins and Oblivion in Victorian Narrative." In *Sex, Politics, and Science in the Nineteenth-Century Novel*, edited by Ruth Bernard Yeazell, 63–87. Baltimore: Johns Hopkins University Press, 1986.

"Behind the Scenes at Vesuvius." *All the Year Round* 9 (1863): 345–48.

Bellamy, Edward. *Looking Backward, 2000–1887*. 1888. New York: Signet, 2000.

Benjamin, Walter. "Theses on the Philosophy of History." In *Illuminations*, edited by Hannah Arendt, translated by Harry Zohn, 255–67. New York: Harcourt, Brace and World, 1968.

Bennett, Tony. *The Birth of the Museum: History, Theory, Politics*. New York: Routledge, 1995.

Berry, William B. N. *Growth of a Prehistoric Time Scale: Based on Organic Evolution*. Palo Alto, CA: Blackwell Scientific Publications, 1987.

Black, Barbara J. *On Exhibit: Victorians and Their Museums.* Charlottesville, VA: University of Virginia Press, 2000.

Black, William Henry. "Observations on the Primitive Site, Extent, and Circumvallation of Roman London." *Archaeologia* 40 (1866): 41–58.

Bohrer, Frederick. *Orientalism and Visual Culture: Imagining Mesopotamia in Nineteenth-Century Europe.* Cambridge: Cambridge University Press, 2003.

Booth, Alison. "Amy Dorrit and Dorothea Brooke: Interpreting the Heroines of History." *Nineteenth-Century Literature* 41 (1986): 190–216.

Borely, Lester, ed. *Celebrating the Life and Times of Hugh Miller.* Cromarty, Scotland: Cromarty Arts Trust and Elphinstone Institute of the University of Aberdeen, 2003.

Bowden, Mark. *Pitt Rivers: The Life and Archaeological Work of Augustus Henry Lane Fox Pitt Rivers.* Cambridge: Cambridge University Press, 1991.

Brand, Vanessa, ed. *The Study of the Past in the Victorian Age.* Oxford: Oxbow Books, 1998.

Brantlinger, Patrick. *Rule of Darkness: British Literature and Imperialism, 1830–1914.* Ithaca, NY: Cornell University Press, 1988.

Bratten, Joel J. "Constancy, Change, and the Dust Mounds of *Our Mutual Friend.*" *Dickens Quarterly* 19 (2002): 23–30.

Brilliant, Richard. *Pompeii AD 79: The Treasure of Rediscovery.* New York: Clarkson N. Potter, an official publication of the American Museum of Natural History, 1979.

Buckland, Francis T. *Curiosities of Natural History.* First Series. London: Richard Bentley and Son, 1893.

Buckley, Jerome H., ed. *Poems of Tennyson.* Boston: Houghton Mifflin, 1958.

———. *Tennyson: The Growth of a Poet.* Cambridge, MA: Harvard University Press, 1960.

———. *The Triumph of Time: A Study of the Victorian Concepts of Time, History, Progress, and Decadence.* Cambridge, MA: Belknap Press of Harvard University Press, 1966.

Bulwer-Lytton, Edward. *The Last Days of Pompeii.* 1834. New York: Fine Editions Press, 1956.

Burstyn, Joan. *Victorian Education and the Ideal of Womanhood.* Rutgers, NJ: Rutgers University Press, 1984.

Campbell, Nancie, compiler, ed. *Tennyson in Lincoln: A Catalogue of the Collections in the Research Centre,* vol. 1. Lincoln, UK: Tennyson Society, 1971.

Cantor, Geoffrey, et al. *Science in the Nineteenth-Century Periodical.* Cambridge: Cambridge University Press, 2004.

Carignan, Michael I. "Fiction as History, or History as Fiction: George Eliot, Hayden White, and Nineteenth-Century Historicism." *Clio* 29 (2000): 395–415.

Carlisle, Janice M. "*Little Dorrit*: Necessary Fictions." *Studies in the Novel* 7 (1975): 195–214.

Ceram, C. W. *Gods, Graves, and Scholars: The Story of Archaeology.* New York: Vintage Books, 1951.

Chambers, Robert. *Vestiges of the Natural History of Creation.* 1844. New York: Wiley and Putnam, 1845.

Chase, Karen, and Michael Levenson. "On the Parapets of Privacy." In *A Companion to Victorian Literature and Culture*, edited by Herbert F. Tucker, 425–37. Malden, MA: Blackwell, 1999.

———. *The Spectacle of Intimacy: A Public Life for the Victorian Family.* Princeton: Princeton University Press, 2000.

Christie, John, and Sally Shuttleworth, eds. *Nature Transfigured: Science and Literature, 1700–1900.* Manchester: Manchester University Press, 1989.

Cohen, Claudine. *The Fate of the Mammoth: Fossils, Myth, and History.* Translated by William Rodarmor. Chicago: University of Chicago Press, 2002.

Cook, E. T., and Alexander Wedderburn, ed. *The Works of John Ruskin, Letters I, 1827–1869*, vol. 36. New York: Longmans, Green, 1909.

Coombes, Annie E. *Reinventing Africa: Museums, Material Culture and Popular Imagination in Late Victorian and Edwardian England.* New Haven: Yale University Press, 1994.

Cotsell, Michael. *Companion to* Our Mutual Friend. London: Allen and Unwin, 1986a.

———. "Mr. Venus Rises from the Counter: Dickens's Taxidermist and His Contribution to *Our Mutual Friend*." *Dickensian* 80 (1984): 105–13.

———. "Politics and Peeling Frescoes: Layard of Nineveh and *Little Dorrit*." *Dickens Studies Annual* 15 (1986b): 181–200.

Crawford, Iain. "'Machinery in Motion': Time in *Little Dorrit*." *Dickensian* 84 (1988): 30–41.

Curl, James Stevens. *Egyptomania—The Egyptian Revival: A Recurring Theme in the History of Taste.* New York: Manchester University Press, 1994.

Daniel, Glyn. *A Hundred and Fifty Years of Archaeology.* Cambridge, MA: Harvard University Press, 1976.

Daniel, Glyn, and Colin Renfrew. *The Idea of Prehistory*. Edinburgh: Edinburgh University Press, 1988.

Darwin, Charles. *The Origin of Species*. London: J. Murray, 1859.

———. *The Voyage of the Beagle*. 1839. New York: Everyman's Library, 1959.

Daubeny, C. G. B., ed. *Fugitive Poems, connected with Natural History and Physical Science*. Oxford: James Parker, 1869.

Dawson, Gowan, Richard Noakes, and Jonathan R. Topham. Introduction. In *Science in the Nineteenth-Century Periodical*, edited by G. N. Cantor, 1–34. Cambridge: Cambridge University Press, 2004.

"A Day at Pompeii." *Harper's New Monthly Magazine* 11 (1855): 721–43.

Dean, Dennis R. *Gideon Mantell and the Discovery of Dinosaurs*. Cambridge: Cambridge University Press, 1999.

———. *Tennyson and Geology, Tennyson Society Monographs: Number Ten*. Lincoln, UK: Tennyson Society, Tennyson Research Centre, 1985.

De la Beche, H[enry] T[homas]. *How to Observe—Geology*. London: Charles Knight, 1835.

Derrida, Jacques. *Of Grammatology*. Corrected edition. Translated by Gayatri Chakravorty Spivak. Baltimore: Johns Hopkins University Press, 1998.

De Vere, Aubrey. "Tennyson's *Princess*." *Edinburgh Review* 90 (1849): 388–433.

Dickens, Charles. *Bleak House*. 1853. Edited by Norman Page. London: Penguin, 1985a.

———. *The Christmas Books*, vol. 1. Edited by Michael Slater. New York: Penguin, 1985b.

———. *Little Dorrit*. 1857. Edited by John Holloway. New York: Penguin, 1967.

———. *Our Mutual Friend*. 1865. Edited by Adrian Poole. New York: Penguin, 1997.

———. *Pictures from Italy*. 1846. New York: Oxford University Press, 1903.

Dillon, Steve. "The Archaeology of Victorian Literature." *Modern Language Quarterly* 54 (1993): 237–61.

Dowling, Linda. "Roman Decadence and Victorian Historiography." *Victorian Studies* 28 (1985): 579–607.

Easson, Angus. "A Novel Scarcely Historical? Time and History in Dickens's *Little Dorrit*." *Essays and Studies* 44 (1991): 27–40.

Edgecombe, Rodney Stenning. "Reading through the Past: Archaeological Conceits and Procedures in *Little Dorrit*." *Yearbook of English Studies* 26 (1996): 65–72.

Eiseley, Loren. *Darwin's Century: Evolution and the Men Who Discovered It.* New York: Anchor Books, 1961.

Elam, Diane. "'Another Day Done and I'm Deeper in Debt': *Little Dorrit* and the Debt of the Everyday." In *Dickens Refigured: Bodies, Desires and Other Histories*, edited by John Schad, 157–77. New York: Manchester University Press, 1996.

Eliot, George. *Middlemarch.* 1871. Edited by David Carroll. Oxford: Oxford University Press, 1996.

Eliot, T. S. *The Waste Land.* 1922. Edited by Valerie Eliot. New York: Harcourt Brace, 1971.

"The English Pompeii." *Littell's Living Age* 3 (1862): 23–26.

Fabian, Johannes. *Time and the Other: How Anthropology Makes Its Object.* New York: Columbia University Press, 1983.

Fielding, K. J. "Dickens's Work with Miss Coutts:—I. 'Nova Scotia Gardens and What Grew There.'" *The Dickensian* 61 (1965): 112–19.

Findlen, Paula. *Possessing Nature: Museums, Collecting, and Scientific Culture in Early Modern Italy.* Berkeley: University of California Press, 1994.

Flint, Kate. "Counter-Historicism, Contact Zones, and Cultural History." *Victorian Literature and Culture* 27, no. 2 (1999): 507–11.

Forde, H. A. *Mary Anning: The Heroine of Lyme Regis.* London: Wells Gardner, Darton, 1925.

Fortey, Richard. *Trilobite! Eyewitness to Evolution.* New York: Alfred A. Knopf, 2000.

Foucault, Michel. "Of Other Spaces." Translated by Jay Miskowiec. *Diacritics* 16, no. 1 (1986): 22–27.

Frank, Katherine, and Steve Dillon. "Description of Darkness: Control and Self Control in Tennyson's Princess." *Victorian Poetry* 33 (1995): 233–55.

Fraser, J. T. *Time, Conflict, and Human Values.* Chicago: University of Illinois Press, 1999.

Fulweiler, Howard W. "'A Dismal Swamp': Darwin, Design, and Evolution in *Our Mutual Friend.*" *Nineteenth-Century Literature* 49, no. 1 (1994): 51–74.

Gaughan, Richard T. "Prospecting for Meaning in *Our Mutual Friend.*" *Dickens Studies Annual* 19 (1990): 231–46.

Gell, Sir William. *Pompeiana: The Topography, Edifices, and Ornaments of Pompeii, the Result of Excavations since 1819.* London: Jennings and Chaplin, 1832.

Gell, Sir William, and J. P. Gandy. *Pompeii: Its Destruction and Re-Discovery with Engravings and Descriptions of the Art and Architecture of Its Inhabitants.* New York: R. Worthington, 1880.

The Geologist. Edited by S. J. Mackie. London: Reynolds, 1858.

Gillispie, Charles Coulston. *Genesis and Geology: The Impact of Scientific Discoveries on Religious Beliefs in the Decades before Darwin.* New York: Harper and Brothers, 1951.

Gilmour, Robin. *The Victorian Period: The Intellectual and Cultural Context of English Literature, 1830–1890.* New York: Longman, 1993.

Gliserman, Susan. "Early Victorian Science Writers and Tennyson's *In Memoriam*: A Study in Cultural Exchange." *Victorian Studies* 18 (1974–75): 277–308, 437–59.

Goldicutt, John. *Specimens of Ancient Decorations from Pompeii.* London: Rodwell and Martin, 1825.

Goldstein, Laurence. "The Impact of Pompeii on the Literary Imagination." *Centennial Review* 23 (1979): 227–41.

Gould, Stephen Jay. *Time's Arrow Time's Cycle: Myth and Metaphor in the Discovery of Geological Time.* Cambridge, MA: Harvard University Press, 1987.

"The Graffiti of Pompeii." *Edinburgh Review* 110 (1859): 411–37.

"Graffiti, or Wall-Scribblings." *Chambers's Journal* 58 (1881): 97–100.

Graham, Ian. *Alfred Maudslay and the Maya.* Norman: University of Oklahoma Press, 2002.

[Gray, Thomas.] *The Vestal, or a Tale of Pompeii.* Boston: Gray and Bowers, 1830.

Grayson, Donald K. *The Establishment of Human Antiquity.* New York: Academic Press, 1983.

Greene, Mott T. *Geology in the Nineteenth Century: Changing Views of a Changing World.* Ithaca, NY: Cornell University Press, 1982.

"Hail Columbia-Square!" *All the Year Round* 7 (1862): 301–6.

Hallam, A. *Great Geological Controversies.* Oxford: Oxford University Press, 1989.

Hamilton, Paul. *Historicism*, 2nd edition. London and New York: Routledge, 2003.

Hardy, Thomas. *A Pair of Blue Eyes.* 1873. Edited by Alan Manford. Oxford: Oxford University Press, 1985.

———. *Tess of the D'Urbervilles.* 1891. Edited by Tim Dolin. London: Penguin Books, 2003.

Harrison, Frederic. "A Pompeii for the Twenty-Ninth Century." *The Nineteenth Century* 28 (1890): 381–91.

Hawker, R[obert]. S[tephen]. *Poems*. Stratton: J. Roberts, 1836.

Haycock, David Boyd. *William Stukeley: Science, Religion and Archaeology in Eighteenth-Century England*. Rochester, NY: Boydell Press, 2002.

Hemans, Felicia Dorothea. "The Image in Lava." In *The New Oxford Book of Romantic Period Verse*, edited by Jerome J. McGann, 720–21. Oxford: Oxford University Press, 1993.

Herbert, Isolde Karen. "'A Strange Diagonal': Ideology and Enclosure in the Framing Sections of *The Princess* and *The Earthly Paradise*." *Victorian Poetry* 29 (1991): 145–59.

Herbert, Sandra. "Between Genesis and Geology: Darwin and Some Contemporaries in the 1820s and 1830s." In *Religion and Irreligion in Victorian Society*, edited by R. W. Davis and R. J. Helmstadter, 68–84. London: Routledge, 1992.

Heringman, Noah. *Romantic Science: The Literary Forms of Natural History*. New York: State University of New York Press, 2003.

Hines, John. *Voices in the Past: English Literature and Archaeology*. Cambridge: Brewer, 2004.

Hollington, Mike. "Time in *Little Dorrit*." In *The English Novel in the Nineteenth Century: Essays on the Literary Mediation of Human Values*, edited by George Goodin, 109–25. Urbana: University of Illinois Press, 1972.

Horne, R. H. "Dust; or Ugliness Redeemed." *Household Words* 1 (1850): 379–84.

Hudson, Kenneth. *A Social History of Archaeology: The British Experience*. London: Macmillan, 1981.

Hunt, Felicity. *Lessons for Life: The Schooling of Girls and Women, 1850–1950*. New York: B. Blackwell, 1987.

Hutton, James. "Theory of the Earth." *Transactions of the Royal Society of Edinburgh* 1 (1788): 209–305.

Huxley, Thomas Henry. "On the Method of Zadig: Retrospective Prophecy as a Function of Science." 1880. Project Gutenberg, 2001. September 21, 2005. <http://www.gutenberg.org/dirs/etext01/1saht10.txt.>

Illustrated London News. "Dinner in the Iguanodon Model, at the Crystal Palace, Sydenham." January 7, 1854.

———. "The Metropolitan Railway and the Fleet Ditch." February 15, 1862.

———. "Roman Villa Discovered in Lower Thames Street." February 5, 1848.

———. "Works in Progress in Threadneedle Street." February 10, 1855.

Jefferies, Richard. *After London; or, Wild England*. 1885. New York: E.P. Dutton, 1906.

Jenkins, Ian. *Archaeologists and Aesthetes in the Sculpture Galleries of the British Museum 1800–1939*. London: British Museum Press, 1992.

Jerrold, Blanchard and Gustav Doré. 1872. *London: A Pilgrimage*. New York: B. Blom, 1968.

Johnson, Edgar, ed. *The Heart of Charles Dickens, as Revealed in His Letters to Angela Burdett-Coutts*. Boston: Little, Brown, 1952.

Johnston, Eileen Tess. "'This Were a Medley': Tennyson's *The Princess*." *ELH* 51, no. 3 (1984): 549–74.

Kendall, May. "The Lay of the Trilobite." 1887. In *Victorian Women Poets*, edited by Angela Leighton and Margaret Reynolds, 630. Oxford: Blackwell, 1995.

Kermode, Frank. *The Sense of an Ending*. Oxford: Oxford University Press, 1966.

Killham, John. *Tennyson and* The Princess*: Reflections of an Age*. London: University of London, Athlone Press, 1958.

Kingsley, Charles. "Tennyson." *Fraser's Magazine* 42 (1850): 245–55.

———. *Town Geology*. London: Strahan, 1872.

Klaver, J. M. I. *Geology and Religious Sentiment: The Effect of Geological Discoveries on English Society and Literature between 1829 and 1859*. New York: Brill, 1997.

Knell, Simon J. *The Culture of English Geology, 1815–1851: A Science Revealed through Its Collecting*. Burlington, VT: Ashgate, 2000.

Kuhn, Thomas S. *The Structure of Scientific Revolutions*. Chicago: University of Chicago Press, 1962.

Lang, Cecil Y., and Edgar F. Shannon, Jr., eds. *The Letters of Alfred Lord Tennyson*, vol. 1, *1821–1850*. Cambridge, MA: Harvard University Press, 1981.

Lang, W. D. "Three Letters by Mary Anning, 'Fossilist' of Lyme." *Proceedings of the Dorset Natural History and Archaeological Society* 66 (1944): 169–73.

Larsen, Janet. "Apocalyptic Style in *Little Dorrit*." *Dickens Quarterly* 1 (1984): 41–49.

"A Last Day at Pompeii." *All the Year Round* 35 (1884): 41–45.

"The Last Days of London." *Punch* 14 (1848): 155.

"Latest News from the Dead." *All the Year Round* 9 (1863): 473–76.

Layard, Austen Henry. *Nineveh and Its Remains*. 1849. Edited by H. W. F. Saggs. London: Routledge and Kegan Paul, 1970.

[———]. "Pompeii." *Quarterly Review* 115 (1864): 312–48.

Leask, Nigel. *Curiosity and the Aesthetics of Travel Writing, 1770–1840*. Oxford: Oxford University Press, 2002.

Levine, George. *Darwin and the Novelists: Patterns of Science in Victorian Fiction*. Cambridge, MA: Harvard University Press, 1988.

Levine, Philippa. *The Amateur and the Professional: Antiquarians, Historians, and Archaeologists in Victorian England, 1838–1886*. London: Cambridge University Press, 1986.

[Lewis, John Delaware]. "The City of Sudden Death." *Household Words* 5 (1852a): 171–76.

[———]. "Preservation in Destruction." *Household Words* 5 (1852b): 280–84.

The Library of Entertaining Knowledge: Pompeii. London: Charles Knight, 1831.

Lightman, Bernard. "'The Voices of Nature': Popularizing Victorian Science." In *Victorian Science in Context*, edited by Bernard Lightman, 187–211. Chicago: University of Chicago Press, 1997.

Lombroso-Ferrero, Gina. *Criminal Man, According to the Classification of Cesare Lombroso*. 1876. Montclair, NJ: Patterson Smith, 1972.

"London in A.D. 2346." *Punch* 11 (1846): 186.

Lyell, Charles. *Lectures on Geology, Delivered at the Broadway Tabernacle*. Edited by Greeley and McElrath. New York: Reported for the *New York Tribune*, 1843.

———. *Life, Letters and Journals of Sir Charles Lyell, Bart.* Edited by Katherine Lyell. London: John Murray, 1881.

———. *Principles of Geology*, vol. 1. 1830. Edited by Martin J. S. Rudwick. Chicago: University of Chicago Press, 1990.

———. *Principles of Geology*, vols. 2–3. 1832–1833. Edited by Martin J. S. Rudwick. New York: Johnson Reprint Corporation, 1969.

———. Review of "Memoir on the Geology of Central France." *Quarterly Review* 36 (1827): 437–83.

Macaulay, Thomas Babington. "Pompeii: A Poem." In *Cambridge Prize Poems*. London: T. and J. Allman, 1828.

———. "Ranke's *History of the Popes.*" *Edinburgh Review* 71 (1840): 227–58.

Malley, Shawn. "Austen Henry Layard and the Periodical Press: Middle Eastern Archaeology and the Excavation of Cultural Identity in Mid-Nineteenth Century Britain." *Victorian Review* 22, no. 2 (1996): 152–70.

———. "Shipping the Bull: Staging Assyria in the British Museum." *Nineteenth-Century Contexts* 26, no. 1 (2004): 1–27.

Manning, Peter J. "Cleansing the Images: Wordsworth, Rome, and the Rise of Historicism." *Texas Studies in Language and Literature* 33 (1991): 271–326.

Manning, Sylvia. "Social Criticism and Textual Subversion in *Little Dorrit.*" *Dickens Studies Annual* 20 (1991): 127–47.

Mantell, Gideon. "Illustrations of the Connexion between Archaeology and Geology." *The Edinburgh New Philosophical Journal* 50 (1851a): 235–54.

———. *The Journal of Gideon Mantell, Surgeon and Geologist.* Edited by Cecil Curwen. London: Oxford University Press, 1940.

———. *The Medals of Creation; or, First Lessons in Geology, and the Study of Organic Remains.* London: Henry G. Bohn, 1844.

———. *Petrifactions and Their Teachings; or, A Hand-Book to the Gallery of Organic Remains of the British Museum.* London: Henry G. Bohn, 1851b.

———. *A Pictorial Atlas of Fossil Remains.* London: Henry G. Bohn, 1850.

———. *The Wonders of Geology; or, A Familiar Exposition of Geological Phenomena.* 4th ed. 1838. London: Henry G. Bohn, 1848.

Marsden, Barry. *The Early Barrow-Diggers.* New Jersey: Noyes Press, 1974.

"Mary Anning, The Fossil Finder." *All the Year Round* 13 (1865): 60–63.

Mattes, Eleanor B. *In Memoriam: The Way of a Soul.* New York: Exposition Press, 1951.

Mayhew, Henry. *London Labour and the London Poor.* 1861–1862. London: Penguin, 1985.

McClintock, Anne. *Imperial Leather: Race, Gender and Sexuality in the Colonial Contest.* New York: Routledge, 1995.

McPhee, John. *Basin and Range.* New York: Farrar, Strauss, and Giroux, 1980.

Meteyard, Eliza. *Hallowed Spots of Ancient London.* London: E. Marlborough, 1862.

Metz, Nancy Aycock. "The Artistic Reclamation of Waste in *Our Mutual Friend.*" *Nineteenth-Century Fiction* 34 (1979): 59–72.

———. "The Blighted Tree and the Book of Fate: Female Models of Story-telling in *Little Dorrit.*" *Dickens Studies Annual* 18 (1989): 221–41.

———. "*Little Dorrit*'s London: Babylon Revisited." *Victorian Studies* 33, no. 3 (1990): 465–86.

———. "Science in *Household Words*: 'The Poetic . . . Passed Into Our Common Life.'" *Victorian Periodicals Newsletter* 11, no. 4 (1978): 121–33.

Middleton, S[tephen]. *Pompeii, a Didactic Poem*. London: Smith, Elder, 1835.

Miller, Edward. *That Noble Cabinet: A History of the British Museum*. Athens: Ohio University Press, 1974.

Miller, J. Hillis. *Charles Dickens: The World of His Novels*. Cambridge, MA: Harvard University Press, 1965.

———. *The Disappearance of God*. Cambridge, MA: Harvard University Press, 1963.

———. *Thomas Hardy: Distance and Desire*. Cambridge, MA: Harvard University Press, 1970.

Millhauser, Milton. *Fire and Ice: The Influence of Science on Tennyson's Poetry*. Lincoln: Tennyson Society, 1971.

———. "The Literary Impact of *Vestiges of Creation*." *Modern Language Quarterly* 17 (1956): 213–26.

———. "Tennyson's *Princess* and *Vestiges*." *PMLA* 69, no. 1 (1954) 337–43.

———. "Tennyson, *Vestiges*, and the Dark Side of Science." *Victorian Newsletter* 35 (1969): 22–25.

Momigliano, Arnaldo. *Studies in Historiography*. London: Wiedenfeld and Nicolson, 1966.

Nead, Lynda. *Victorian Babylon: People, Streets and Images in Nineteenth-Century London*. New Haven: Yale University Press, 2000.

Newsom, Robert. "Administrative." In *A Companion to Victorian Literature and Culture*, edited by Herbert F. Tucker, 212–24. Malden, MA: Blackwell, 1999.

Newton, Charles. "On the Study of Archaeology." *The Archaeological Journal* 8 (1851): 1–26.

Nordau, Max. *Degeneration*. 1892. New York: H. Fertig, 1968.

Olsen, Donald J. *The Growth of Victorian London*. New York: Holmes and Meier Publishers, 1976.

Park, Hyungji. "'Going to Wake Up Egypt': Exhibiting Empire in *Edwin Drood*." *Victorian Literature and Culture* (2002): 529–50.

Pascoe, Judith. *The Hummingbird Cabinet: A Rare and Curious History of Romantic Collectors.* Ithaca: Cornell University Press, 2006.

Patten, Robert L. "Dickens Time and Again." *Dickens Studies Annual* 2 (1972): 162–92.

Pearce, Susan M. "Giovanni Battista Belzoni's Exhibition of the Reconstructed Tomb of Pharoah Seti I in 1821." *Journal of the History of Collections* 12, no. 1 (2000): 109–25.

Pearson, Mike, and Michael Shanks. *Theatre/Archaeology.* New York: Routledge, 2001.

Piggott, Stuart. *Ancient Britons and the Antiquarian Imagination.* New York: Thames and Hudson, 1989.

———. *Ruins in a Landscape: Essays in Antiquarianism.* Edinburgh: University Press of Edinburgh, 1976.

———. *William Stukeley: An Eighteenth-Century Antiquary.* New York: Thames and Hudson, 1985.

"Pompeii by Torchlight." *New Monthly Magazine* 43 (1835): 64–69.

Poovey, Mary. *Making a Social Body: British Cultural Formation, 1830–1864.* Chicago: University of Chicago Press, 1995.

Porter, Roy. "Charles Lyell and the Principles of the History of Geology." *British Journal for the History of Science* 9 (1976): 91–103.

Poulet, Georges. *Studies in Human Time.* Translated by Elliott Coleman. Baltimore: Johns Hopkins University Press, 1950.

Price, John Edward. "[An Index of] Roman Remains in London." *Archaeological Review* 1 (1888): 274–79, 355–61.

Radice, Betty, trans. *The Letters of Pliny the Younger.* Harmondsworth: Penguin, 1969.

Richards, Thomas. *The Commodity Culture of Victorian England: Advertising and Spectacle, 1851–1914.* Stanford, CA: Stanford University Press, 1990.

Ricoeur, Paul. "The Human Experience of Time and Narrative." In *A Ricoeur Reader: Reflection and Imagination,* edited by Mario J. Valdés, 99–116. Toronto and Buffalo: University of Toronto Press, 1991.

———. *Time and Narrative.* Translated by Kathleen Blarney and David Pellauer, vol. 3. Chicago: University of Chicago Press, 1985.

Roach Smith, Charles. "Catalogue of the Museum of London Antiquities (collected by, and the property of, Charles Roach Smith)." Printed for the subscribers only, 1854.

———. *Illustrations of Roman London*. London: Printed for the subscribers only, 1859.

———. "Observations on the Roman Remains Found in Various Parts of London." *Archaeologia* 27 (1838): 140–52.

———. "Observation on the Roman Remains Recently Found in London." *Archaeologia* 29 (1842): 145–67, 267–74.

———. "On Some Roman Bronzes Discovered in the Bed of the Thames." *Archaeologia* 28 (1840): 38–46.

———. *Retrospections, Social and Archaeological*, 2 vols. London: George Bell and Sons, 1883, 1886.

———. "Roman London." *Archaeological Journal* 1 (1844): 108–17.

"Roman London." *Bentley's Miscellany* 59 (1866): 364–68.

Roopnaraine, R. Rupert. "Time and the Circle in *Little Dorrit*." *Dickens Studies Annual* 3 (1974): 54–76.

Rossetti, Dante Gabriel. "The Burden of Nineveh." 1850. In *Victorian Literature and Culture*, edited by Dorothy Mermin and Herbert Tucker, 804. New York: Harcourt College Publishers, 2002.

Rossi, Paolo. *The Dark Abyss of Time: The History of the Earth and the History of Nations from Hooke to Vico*. Translated by Lydia G. Cochrane. Chicago: University of Chicago Press, 1984.

Roth, Michael S., with Claire Lyons and Charles Merewether. *Irresistible Decay: Ruins Reclaimed*. Los Angeles, CA: Getty Research Institute for the History of Art and the Humanities, 1997.

Rudwick, Martin J. S. *Bursting the Limits of Time: The Reconstruction of Geohistory in the Age of Revolution*. Chicago: University of Chicago Press, 2005.

———. *The Great Devonian Controversy*. Chicago: University of Chicago Press, 1985.

———. Introduction. In *Principles of Geology*, vii–lviii. Chicago: University of Chicago Press, 1990.

———. *The Meaning of Fossils: Episodes in the History of Palaeontology*. 2nd ed. New York: Neale Watson Academic Publications, 1976.

———. *Scenes from Deep Time: Early Pictorial Representations of the Prehistoric World*. Chicago: University of Chicago Press, 1992.

———. "The Strategy of Lyell's *Principles of Geology*." *Isis* 61 (1970): 5–33.

"The Ruined City." *All the Year Round* 2 (1860): 249.

Rupke, Nicolaas A. *The Great Chain of History: William Buckland and the English School of Geology (1814–1849)*. Oxford: Clarendon, 1983.

Ruskin, John. *Works*. Edited by T. Cook and Alexander Wedderburn, 39 vols. London: George Allen, 1903–1912.

Scharf, Sir George. *The Pompeian Court in the Crystal Palace*. London: Crystal Palace Library, and Bradbury and Evans, 1854.

Scholes, Robert, and Robert Kellogg. *The Nature of Narrative*. Oxford: Oxford University Press, 1966.

Schwarzbach, F. S. *Dickens and the City*. London: Athlone Press, 1979.

Scrope, George Poulett. *Memoir on the Geology of Central France*. London: Longman, Rees, Orme, Brown, and Green, 1827.

———. Review of *Principles of Geology* by Charles Lyell. *Quarterly Review* 43 (1830): 411–69.

Secord, James A. *Controversy in Victorian Geology: The Cambrian-Silurian Dispute*. Princeton, NJ: Princeton University Press, 1986.

———. *A Victorian Sensation: The Extraordinary Publication, Reception, and Secret Authorship of Vestiges of the Natural History of Creation*. Chicago: University of Chicago Press, 2000.

Sedgwick, Eve Kosofsky. "'Tennyson's Princess: One Bride for Seven Brothers." In *Tennyson*, edited by Rebecca Stott, 181–96. London: Longman, 1996.

Shanks, Michael, and Christopher Tilley. *Re-Constructing Archaeology*. New York: Routledge, 1992.

Shelley, Percy. "Ozymandias." 1818. In *Shelley's Poetry and Prose*, edited by Donald H. Reiman, 103. New York: Norton, 1977.

Sicher, Efraim. *Rereading the City Rereading Dickens: Representation, the Novel, and Urban Realism*. New York: AMS Press, 2003.

Smith, Horatio. "A New Nightmare." In *Fugitive Poems, connected with Natural History and Physical Science*, edited by C. G. B. Daubeny, 123–26. Oxford: James Parker, 1869.

Smith, Jonathan. "Darwin's Barnacles, Dickens's *Little Dorrit*, and the Social Uses of Victorian Seaside Studies." *LIT: Literature, Interpretation, Theory* 10 (2000): 327–47.

———. "Heat and Modern Thought—Charles Dickens: The Forces of Nature in *Our Mutual Friend*." *Victorian Literature and Culture* 23 (1995): 37–69.

Snowden, Eleanor. *The Moorish Queen: A Record of Pompeii; and Other Poems*. London: Longman, 1831.

Stafford, Fiona J. *The Last of the Race: The Growth of a Myth from Milton to Darwin*. Oxford: Clarendon Press, 1994.

Starke, Mariana. *Travels in Europe*. London: John Murray, 1832.

Stauffer, Andrew M. "Dante Gabriel Rossetti and the Burdens of Nineveh." *Victorian Literature and Culture* 33 (2005): 369–94.

———. "Ruins of Paper: Dickens' Necropolitan London." Work in progress.

Stone, Marjorie. "Genre Subversion and Gender Inversion: *The Princess* and *Aurora Leigh*." *Victorian Poetry* 25 (1987): 101–26.

Sucksmith, Harvey Peter. "The Dust-Heaps in *Our Mutual Friend*." *Essays in Criticism* 23 (1973): 206–12.

Taylor, Michael A., and Hugh S. Torrens. "Saleswoman to a New Science." *Proceedings of the Dorset Natural History and Archaeological Society* 108 (1987): 135–48.

Tennant, James. *Catalogue of Fossils, Found in the British Isles, Forming the Private Collection of James Tennant, F.G.S.* London: J. Tennant, 1838.

Tennyson, Alfred. *The Poems of Tennyson in Three Volumes*, vol 2. Edited by Christopher Ricks. Berkeley: University of California Press, 1987.

———. *Tennyson*. 2nd ed. Edited by Christopher Ricks. Berkeley: University of California Press, 1989.

Thackeray, William Makepeace. *The Newcomes*. 1853. Edited by David Pascoe. London: Penguin, 1996.

Tickell, Crispin. *Mary Anning of Lyme Regis*. Lyme Regis, UK: Lyme Regis Philpot Museum, 1996.

Times (London). "The Excavations at Pompeii." February 18, 1845a.

———. "Pompeii." October 20, 1837.

———. "Pompeii." July 18, 1845.

———. "Roman London." October 19, 1859.

———. "Ruins of Pompeii." January 28, 1837.

———. "State of the Thames." June 17, 1858.

Tite, William, Esq. *A Descriptive Catalogue of the Antiquities Found in the Excavations at the New Royal Exchange, Preserved in the Museum of the Corporation of London*. Printed for the use of the members of the Corporation of the City of London, 1848.

Tomko, Michael. "Varieties of Geological Experience: Religion, Body, and Spirit in Tennyson's *In Memoriam* and Lyell's *Principles of Geology*." *Victorian Poetry* 42, no. 2 (1995): 257–84.

Torrens, H[ugh]. S. "Geology and the Natural Sciences: Some Contributions to Archaeology in Britain 1780–1850." In *The Study of the Past in the*

Victorian Age, edited by Vanessa Brand, 35–59. Oxford: Oxbow Books, 1998.

Toulmin, Stephen, and June Goodfield. *The Discovery of Time*. Chicago: University of Chicago Press, 1965.

Trigger, Bruce G. *A History of Archaeological Thought*. Cambridge: Cambridge University Press, 1989.

Tucker, Herbert F. "The Fix of Form: An Open Letter." *Victorian Literature and Culture* 27, no. 2 (1999): 531–35.

———. "Of Monuments and Moments: Spacetime in Nineteenth-Century Poetry." *Modern Language Quarterly* 58, no. 3 (1997): 269–97.

Vance, Norman. *The Victorians and Ancient Rome*. Cambridge, MA: Blackwell, 1997.

Van Riper, A. Bowdoin. *Men among the Mammoths: Victorian Science and the Discovery of Human Prehistory*. Chicago: University of Chicago Press, 1993.

"What Is to Become of Us?" *Household Words* 5 (1852): 352–56.

White, Hayden. *Metahistory: The Historical Imagination in Nineteenth-Century Europe*. Baltimore: Johns Hopkins University Press, 1973.

———. *Tropics of Discourse: Essays in Cultural Criticism*. Baltimore: Johns Hopkins University Press, 1978.

Wickens, G. Glen. "The Two Sides of Early Victorian Science and the Unity of *The Princess*." *Victorian Studies* 23 (1980): 369–88.

Williams, Raymond. *Culture and Society: 1780–1950*. New York: Columbia University Press, 1958.

Wilson, David. *The British Museum: A History*. London: British Museum Press, 2002.

Wilson, Leonard G. *Charles Lyell, the Years to 1841: The Revolution in Geology*. New Haven: Yale University Press, 1972.

Winchester, Simon. *The Map that Changed the World: William Smith and the Birth of Modern Geology*. New York: HarperCollins, 2001.

Woodring, Carl. "The Burden of Nineveh." *Victorian Newsletter* 63 (1983): 12–14.

Wright, Thomas. "Antiquarianism in England." *Edinburgh Review* 86 (1847): 163–74.

———. "The Romans in Britain." *Edinburgh Review* 94 (1851): 89–103.

Yanni, Carla. *Nature's Museums: Victorian Science and the Architecture of Display*. Baltimore: Johns Hopkins University Press, 1999.

Index